THE ART OF MORE

MICHAEL BROOKS

THE ART OF MORE

How Mathematics Created Civilization

Pantheon Books, New York

Library of Congress Cataloging-in-Publication Data
Name: Brooks, Michael, [date] author.
Title: The art of more : how mathematics created civilization / Michael Brooks.
Description: First American edition. New York : Pantheon Books, 2022.
Includes bibliographical references.
Identifiers: LCCN 2021024094 (print) | LCCN 2021024095 (ebook) |
ISBN 9781524748999 (hardcover) | ISBN 9781524749002 (ebook)
Subjects: LCSH: Mathematics and civilization. Mathematics—History.
Classification: LCC CB455 .B76 2022 (print) | LCC CB455 (ebook) |
DDC 909—dc23
LC record available at https://lccn.loc.gov/2021024094
LC ebook record available at https://lccn.loc.gov/2021024095

www.pantheonbooks.com

Jacket image: Interior of Hagia Sophia, Istanbul, Turkey, 1852/
Science History Images/Alamy
Jacket design by Jenny Carrow

Contents

Author's note

All are welcome here, whether you love mathematics, have always hated it, or just wish you understood it better. There's a whole rainbow of experiences with this subject and, right from the start, I wanted this book to be accessible to every part of the spectrum. To this end, I've kept things as straightforward as possible, but I've occasionally thought it was worth investing a tiny bit of effort to properly understand something. That means there are bits of actual maths in here: some graphs, equations, and calculations that I'll walk you through gently. But if any of it bothers you, and you don't feel like being bothered, just skip that bit. Life's short enough already.

THE ART OF MORE

Introduction

Why our skill with numbers is the greatest human achievement of all

In June 1992, the American psychologist Peter Gordon travelled to a village of leaf-thatched houses on the banks of the Maici River in Amazonia, Brazil.[1] He was there to meet his friend Daniel Everett, who was living as a Christian missionary among the remote and isolated Pirahã people. Everett had told Gordon that the Pirahã have a somewhat relaxed attitude to numbers: essentially, they just don't bother with them. Intrigued, Gordon had come to find out more.

Using the clutch of AA batteries he had brought with him to the jungle, Gordon set up an experiment. He laid some out in a line and asked Pirahã villagers to create another line next to it with the same number of batteries. One, two, or three batteries was not a challenge. But the villagers struggled to correctly match a line of four, five, or six. Ten proved almost impossible. They had the same problem when asked to reproduce marks made on a piece of paper. If there were one or two marks they could do it, but six was the most anyone managed. The Pirahã, it seemed to Gordon, didn't have any kind of handle on numbers — probably because they didn't need to. Their way of life meant

1

that their brains had never had any reason to form number concepts.

To most of us, it comes as a surprise to learn that people can happily get by without numbers. That's because we unconsciously appreciate that numbers are deeply embedded in our daily lives. What we don't appreciate, until it is brought to our attention, is that our way of life, our institutions and our infrastructures were built on numbers. Whether we're talking about business, housing, medicine, politics, warfare, farming, art, travel, science, or technology, almost every aspect of our existence is built on mathematical foundations. And that is all the more astonishing when you appreciate that mathematics didn't *have* to happen.

When it comes to a natural ability with numbers, we're no better off than many other species.[2] Humans are born only with what is now known as an 'approximate number sense'.[3] This means that, in its raw state, your brain doesn't bother specifying when there are more than three of something. So, when a human baby sees four apples, the sight is logged as 'many' or 'more'. Our natural counting system is '1, 2, 3, more'. The brains of rats, chimpanzees, birds and monkeys also use an approximate number system. Reward a rat for pushing a lever five times, and it will occasionally return to the apparatus, performing a varying number of pushes close to five in the hopes of a treat. People have managed to teach chimps to do more sophisticated number-related tasks — remembering sequences of numbers, for example — and they can sometimes be better at this than untrained adult humans. But the training requires rewards: chimps don't start to do maths for the fun of it. And neither did you; you learned to count because of cultural pressures. Those pressures came from an interesting place: a deeply ingrained cultural wisdom that tells us that mathematics matters.

The Tudor mathematician and mystic John Dee called mathematics 'a strange participation between things supernatural, immortal, intellectual, simple, and indivisible and things natural, mortal, sensible, compounded, and divisible'.[4] This might seem like mumbo-jumbo, but mathematics *is* supernatural, in that we have used it to go beyond the natural. Developing maths allowed us to dissect and dismantle nature's patterns and symmetries and, like gods, recast them in ways that serve our interest. Through maths, we shape the world around us to give ourselves a better experience of being human. The first leap was to count to four, and eventually we found ourselves establishing civilizations. Once our brains are schooled in the art of 'more', they become able to cope with complicated abstractions. They grow comfortable in a world where numbers can be applied not just to things that need counting, but to shapes, points, lines and angles — geometry, in other words. This gives us the ability to reimagine — on paper, on a wooden sphere, or just in our minds — a huge and complicated object like the Earth, say, and to navigate our way around it. We can also reimagine numbers — the ones we know and the ones we don't — as symbols that we can manipulate to control and re-engineer the world, performing astounding feats of ordering, optimisation, and transportation. That's algebra, in case you were wondering. We can even do calculations that predict the future that will result from the change that is happening around us. We call these calculus, and they enable us to realise a range of human aspirations, from free market capitalism to moon landings.

We learn this maths — or we are supposed to — early in our lives. In school, we are assured that maths is an essential skill; a passport for success; something that we have to pick up. And so we obediently, though often reluctantly, gather the tools of maths and do our best to learn how to use them. Some enjoy it; most don't. And then, at some point, almost every one of us gives up.

Few of us will learn any more maths after that moment. Over the ensuing years, our hard-won skills wither away and we are left with only the basics at our fingertips. Without technological assistance — such as our mobile phone's calculator, an essential tool for dividing up restaurant bills these days — we find ourselves able to reliably add and subtract only relatively small numbers, and maybe to multiply and divide a little. The rest is lost. We might even become 'maths-phobic', actively avoiding any encounter with numbers. Or we might just think of mathematics as beyond our grasp; something that's 'not for me'.

If that's you, I hope this book will change your mind. The extraordinary achievement that is mathematics belongs to everyone, no matter how good they are — or aren't — with numbers. We all benefit from the way the human brain has put maths to work over millennia, and we all have the right to engage with it, whatever our educational achievements. Why shouldn't you be able to see how Newton's calculus is as beautiful as the Taj Mahal, and why there is as much beauty in the Babylonians' algebra as there once was in their Hanging Gardens? And the proper appreciation of mathematics is not just about some equivalence to what we traditionally see as beauty; it's about seeing how we built the things we value as beautiful. Wherever we look in art or architecture, whether it's one of Vermeer's paintings or Istanbul's majestic Hagia Sophia, we find mathematics facilitating its creation. This influence goes beyond aesthetic issues; the human story is itself inextricably interwoven with mathematics. Columbus's journey to the Americas relied on understanding the properties of triangles, and the modern corporate world is a consequence of what a grasp of numbers makes possible. Mathematics provides the sculptor's chisel that shaped the Renaissance and the ammunition that has engendered centuries of military success. It is the interpreter that allowed people with no common language to establish mutually beneficial trade and the fuel that took humans to the moon. It is the spark that electrified the world at the beginning of the 20th century, and the power behind every throne in the ancient world. No wonder King Shulgi of Ur was worshipped for his mathematical skills four millennia ago.

I learned none of this at school. I did learn how to pass all my maths exams, and sometimes how to apply that maths in order to work out the acceleration of a car or the force required to send a rocket into orbit. But I never learned what mathematics had done for us as a species, or how we came to invent it. It's not too late, though. We can still find joy and meaning in maths, even if decades have passed since we gave up on learning its technicalities.

I still remember where and when I hit my mathematical limit: it was October 1987, and I was in a lecture hall at the University of Sussex in the south of England, having just started my undergraduate physics degree. I don't remember the exact subject, but this lecture was the first in a course on advanced mathematical techniques. The topic just felt too hard for me, and the course was optional, so I walked out. Your story will be different, but at some point we all left our last maths class. Fortunately, the door never really closed behind us. So let's go back in.

Chapter 1

ARITHMETIC

How we founded civilization

Humans didn't evolve with a compulsion to count. But after we invented numbers and arithmetic, we eventually became reliant upon them. Numbers enabled people to govern, tax, and trade with each other, opening up the possibility of living in large interdependent communities. Eventually, arithmetic and its creations — fractions, negative numbers and the concept of zero — became the driving force behind economic and political success: those who can crunch the numbers are those that decide the future of workers, of nations, and even of the planet. And it all started with a mental leap to the number 4.

In the first half of the 15th century, the Medici Bank was the toast of Florence and the envy of Europe.[1] The secret of its success was simple: its chief accountant, Giovanni Benci, was an enthusiast for bookkeeping and a stickler for protocol. He audited the accounts of all the bank's branches every year, checking on the status of debtors and the likelihood of payment defaults. If you managed one of the bank's branches, and your accounts didn't add up, Benci would call you in and tear you apart. And then, in 1455, Benci died and everything fell apart.

The Medici Bank's employees were suddenly free of Benci's prudence, and began promising far too generous a return to depositors, akin to a modern bank guaranteeing a 10 per cent return on any investment. The need to find the money for those guaranteed interest payments led to a toxic lending policy. The bank offered loans at exorbitant interest rates and, desperate to finance their wars, European kings and noblemen took up the Medici's offers with no intention of paying their debts. The bank had no way of enforcing repayments, and so the money was lost. Meanwhile, the partners in the bank cast their eyes over books that were inflated by the promise of these never-to-be-seen payments, and took the non-existent profits out of the business for their own private spending. Their extravagant lifestyles ran riot, draining the bank of cash. In 1478, the Medici bank began to collapse. Faced with personal ruin, Lorenzo de' Medici, great-grandson of the bank's founder, bailed himself out by raiding public funds. The Florentine public was outraged, and stormed the Medici palace in 1494, setting fire to all its banking records. A century-long domination of Europe's cultural, political, and financial capital went up in smoke.

History's next demonstration of the world-changing power of accounting came with the French Revolution. We can trace its eruption to the sacking of accountant Jacques Necker, who had been trying to fix France's broken financial system and reduce its crippling national debt. In the process, he had exposed the profligate indulgence of the French royal court. Eventually Necker's interference was too much for the ruling classes, who were losing money hand over fist in his reforms. Necker lost his job as finance minister — but gained a loyal and dangerous band of admirers.

The historian François Mignet describes the revolution's inciting moment: the hotheaded Camille Desmoulins stands on a table, pistol in hand.[2] 'Citizens! There is no time to lose!' the young rebel cries. Necker's dismissal, Desmoulins says, is an insult and a threat to every patriotic citizen of France. 'One resource is left; to take arms!' At this rallying cry, crowds rush into the streets. On their shoulders they carry busts of the sacked accountant. Mignon tells us: 'Every crisis requires a leader, whose name becomes the standard of his party; while the assembly contended with the court, that leader was Necker.'

Necker's crusade was focused on something we rarely conceive of as revolutionary: he wanted to balance the books. Necker had pointed out that the English parliament published all its accounts and England's finances were in a healthy state, despite heavy borrowing to finance wars abroad. He was determined that France should achieve the same transparency. Balanced books, Necker said, were the basis of moral, prosperous, happy, and powerful government. And so he attempted to streamline the French government's sprawling array of ledgers into a single account based on books that he would audit himself. The idea was not popular among those in power, but extremely popular among those who were not. And so, as historian Jacob Soll has put it, 'The French Revolution would begin, in part, as a fight about accountability and accurate numbers in government.'[3]

It's not only France that envied foreign financial systems; the pillars of the United States economy — tax revenues, the dollar and the central bank — were copied principally from Dutch and English banking practices. At the time, America had no banks, and was drowning in debt. Banks, said Alexander Hamilton in 1781, were 'the happiest engines that ever were invented for advancing trade'.[4] Hamilton argued that freedom from British rule would come from understanding and controlling the accounts. ''Tis by introducing order into our finances — by restoring public credit — not by gaining battles, that we are finally to gain our object,' he said. 'Great Britain is indebted for the immense efforts she has been able to make in so many illustrious and successful wars essentially to that vast fabric of credit raised on this foundation. 'Tis by this alone she now menaces our independence.'

In his role as first secretary of the treasury, Hamilton put all necessary measures in place and lifted the nascent United States out of the mire of bankruptcy. By 1803, Hamilton's financial nous had enabled the US to raise enough Treasury bonds to purchase the Louisiana Territory from France, doubling the size of America. You might enjoy the musical *Hamilton* as a celebration of one of America's founding fathers, but economic historians enjoy it as a celebration of fiscal prudence. And mathematicians see it as testimony to the power that comes from mastering numbers.

Learning to Count

We shouldn't take mathematics for granted. The modern human — *Homo sapiens*, the 'wise man' — has been around for 300,000 years, and we have found human-created artefacts that are at least 100,000 years old. But our oldest reliable record of human counting is somewhere around 20,000 years old. The markings laid out on the surface of the Ishango Bone, discovered in the Ishango region of what is now known as the Democratic Republic of Congo, are a series of long notches that are grouped into three columns, each of which is subdivided into sets. Though we can't know anything for sure, it doesn't seem too much of a stretch to suppose that a single stroke designates an occurrence of 'one'. Two strokes is 'two' and, well, you get the idea. Taken as a whole, the notches look like a tally system for counting lunar cycles.[5]

The relatively recent creation of this bone suggests that counting is a late-blooming skill, not an inevitable result of intelligence. The brain inside your head is largely the same as the one inside the skull of the first *Homo sapiens*, and it seems that for most of our species' history, this wise man did not bother with numbers at all.

Once we did get to grips with numbers, however, the advantage was clear. This is why you probably don't even remember learning to count. Counting is such a valued skill in most human cultures that you would have started before you began to lay down permanent memories.

And I'm willing to bet that you learned to count using your fingers.[6]

The first time I ever really thought about finger-counting — apart from in embarrassment when I realised I was doing it in public, in a supermarket, as I counted off that night's dinner party guests — was when I saw Quentin Tarantino's riotous war movie *Inglourious Basterds*. During a scene in a basement bar, a British character is pretending to be German. He indicates to the barman that he wants three glasses by holding up his index, middle, and ring fingers. The German officer with whom he is sharing a table knows immediately that his drinking partner is a fraud. 'You've just given yourself away, Captain,' he says.

Germans use the thumb for 'one', so a German would have ordered three glasses using the thumb and the first two fingers.[7] In Asia, people finger-count differently. My friend Sonali, who grew up in India, learned to count using the individual segments of her fingers. Merchants in the Indian state of Maharashtra do it differently again.[8] They start with the thumb, like the Germans, but when they get to five, they raise the thumb of the other hand — usually the right — to indicate one 'five' has been reached. The left fist closes again, and the thumb comes out to indicate 'six'.

Imagine doing business with a Maharashtran merchant. At first you would probably be confused, but it wouldn't take you long to figure out, with no language at all, how much you were being asked to pay. Thanks to finger-counting, you can carry out commercial trade with no common written or spoken language. All you need is for both sides to know what currency you're talking about, and to appreciate the meaning of numbers as they rise from 1 into the hundreds and thousands.

This is why learning finger signs was an essential part of education for almost all members of ancient societies. Even the most isolated communities would barter with passing traders with whom they might have no common language. In his 4th-century BC writings, Aristophanes mentions finger-counting as being a common practice of ancient Greece and Persia. The Roman writer Quintillian talked about the shame that would be heaped upon a lawyer who hesitated over his finger signs for numbers. Aztec paintings depict men using finger signs, and in medieval Europe, finger-counting was so ubiquitous that Luca Pacioli's 1494 acclaimed mathematics textbook *Summa de Arithmetica, Geometrica, Proportiono e Proportionalita* contained a complete illustrated guide to the art. Even as late as the 18th century, the German adventurer Carsten Niebuhr describes Asian market traders conducting covert negotiations by grasping each other's fingers and thumbs in various configurations. To keep their business to themselves, they would do this with their hands hidden inside voluminous sleeves or under a large piece of cloth draped over their wrists.

Because the means of signifying numbers has always varied from culture to culture, students of business had to learn their hand gestures carefully. Poets and teachers created rhymes and aphorisms to help with this, such as this effort from the ancient Arab world. 'Khalid left with a fortune of 90 dirhams, and when he came back he had only a third of it left.' Though it doesn't sound helpful to us, the Arabian finger sign for 90 was an index finger curled tightly against the base of the thumb. One-third of 90 is 30, the sign for which was a much broader circle, with the tip of the index finger held against the tip of the thumb. The implication is that Khalid has been sodomised as well as robbed. I suspect you will now remember these ancient signs for 90 and 30 for the rest of your life.

The reason finger signs are so ubiquitous has a lot to do with the reason that humans became good with numbers, once we realised their value. It's this: over the first five years of your life, through play, experimentation and stimulation, your brain develops something called finger sense, or 'gnosis'. This is the ability to treat and sense each digit separately. After a while, your brain begins to hold an internal representation of your fingers, and this representation is used to help when you start to deal with numbers.[9] The beauty of fingers is that they can be seen, felt and moved. They come in two collections of five units, each of which can be put into different configurations of flexion. If you were to put together a tool for assigning a concept of 'how many?' to a group of objects in front of you, you would struggle to beat your own fingers.

Brain scans show that when most of us are presented with mathematical tasks such as subtracting one number from another, the area of the brain that deals with inputs from the fingers steps up to the plate. If the numbers involved are big, the activation of those brain circuits is even clearer. Interestingly, if you're particularly good at subtraction, your brain's finger circuits don't get quite so active: they're barely breaking sweat, in other words. But it's also worth noting that if you weren't encouraged to use your fingers in play as a child — especially when singing counting songs such as 'One, two, buckle my shoe', you may never have really 'got' numbers.[10] Numbers just won't be represented in your brain in the same way that they are for other people. That's one reason some people struggle with maths.

Once you have numbers at your fingertips, it might seem obvious that the next step is to start writing them down. But if we didn't have to start using numbers, we certainly didn't have to start writing them down. After all, when trade was in the moment, involving face-to-face bargaining and immediate transfer of goods or services, there was no need to keep tabs on the transactions. So what made us develop written numerics? By writing numbers down, we could formulate predictions about celestial events that might have religious relevance — new moons or solar eclipses, say. Or we could create inventories of stock and prices paid, and document promises to buy and sell at some point in the future. Writing numbers probably started as a religious practice but it also allowed us to take trade to the next level. Whatever its origins, it led directly to the prosperity we enjoy today.

The Accounting Revolution

We can't really know who the first people to keep records of numbers were; it may be that the Ishango Bone was notched a long time after humankind's mathematical journey began. We do know two things for sure, however. The first is that there have been myriad forms of numerical notation, starting with notched bones and moving into Incan knots, Babylonian marks on clay, Egyptian ink on papyrus, and eventually the 20th century's electrical voltages inside a microchip. The second is that this new ability to keep financial accounts was revolutionary. You might not think of accountancy as anything other than a chore you're glad someone else can do for you, but its invention shifted human culture on its axis.

Our earliest evidence of commercial accounting comes from around 4,000 years ago, when Mesopotamian traders began making records of agreements to sell sheep. Each agreement was represented by a clay ball. The balls were sealed inside a hollow sphere, which was marked with the number of balls it contained, then baked so that the record could not be altered. It was an insurance against the misremembering — deliberate or otherwise — of what had been agreed.

That system evolved into a simpler record: marks baked onto the surface of a clay tablet. Now it was easy to see what had been agreed, bought, sold, or paid. And by this time, humans were already starting to recognise that manipulating numbers could bring more than trade: it could also bring power.

In 2074 BC, in the region of the world we now call south-west Iran, King Shulgi of Ur introduced what scholars have termed 'the first mathematical state'.[11] Shulgi began with a military reform, and followed it up with an administrative reform. This required the scribes of Ur to create complex accounts of everything in the kingdom. The overseers of the working population of Ur left us records of hours worked, illnesses, absences, and the output of loaned and borrowed slaves. If they were not able to show that they had pressed 30 days' worth of work from each of their workers in each month (regardless of how many days were in the month), the deficit would have to be paid to the state. If the overseer scribe died in deficit, the debt passed to his family. King Shulgi's accounting system was designed around a surprising principle: it should make it as easy as possible to detect attempts to defraud the state. Auditing, it turns out, is the true cradle of civilisation.

If Ur was the first mathematical state, Shulgi was the first mathematical god. He declared himself divine in the twenty-third year of his reign. From this time on, his subjects were instructed to worship him and praise his attributes — and in particular his artistry with numbers. We have records of the hymns that were sung in Shulgi's praise; one of his divine attributes was, apparently, his extensive mathematical training in the 'tablet-house', where he learned addition, subtraction, counting, and accounting.

Such were the advantages of putting mathematics at the centre of Shulgi's state that, within a generation, mathematics became the highest art in the land, an essential component of a scribe's training. By the turn of the second millennium BC, a fully qualified scribe would be able to read and write in Sumerian and Babylonian, and know about music and mathematics. The mathematics in question was not the utilitarian number-wrangling of the accountants, but the manipulation of numbers to do extremely difficult — and ostensibly useless — calculations. Essentially, it involved solving riddles such as 'I have added together the perimeter, the diameter and the area of a circle, and the outcome was 115' — the scribe's job was to find the radius.[12] This was maths for maths' sake, and it was considered one of the 'virtues'. Only with mathematical prowess in place could an educated scribe consider himself a master of *nam-lú-ulu*, Sumerian for 'the condition of being human'. In other words, mathematics education first found its place within the curriculum of the humanities.

No wonder, then, that we have found many tens of thousands of ancient clay tablets detailing more than just accounts. Many were used as mathematical aids: multiplication tables, division aids, lists of the squares of numbers (the result of multiplying a number by itself) and the reverse — square roots. We have clay records of how to deal with fractions and algebra, and geometric tools such as an approximate value for pi and the square root of 2. We'll get to the importance of those tools and techniques in later chapters; suffice it to say that, when what we term civilisation began, numbers lay at the heart of society.

With reliable numbers came exceptional power. Thanks, in part at least, to his understanding of the usefulness of mathematics, the reach of Shulgi's kingdom was unprecedented. He finished his father's construction, the Great Ziggurat of Ur, built an extensive road network and oversaw an expanding empire of trade with Arab and Indus communities. All this was possible not because some mathematics had been invented, but because it had been *implemented* — for political purposes. And the same soon became true elsewhere.

We are perhaps over-obsessed with Babylonian and Sumerian mathematical ingenuity, simply because their use of clay for recording daily life has left us with a set of easily accessible artefacts. Societies that used oral traditions are under-represented in our story of how mathematics has always been woven into the fabric of any civilisation. Take the Akan people of West Africa, for instance. In pre-colonial times, they operated a sophisticated mathematical system for weighing gold used in trade. It worked in two strands: one was for working with the Arab and Portuguese systems of weights; the other corresponded to Dutch and English measures. The researchers who finally pieced together its workings from artefacts held in museums around the world suggest that it was so breathtakingly complex that it should be given UNESCO World Heritage status.[13]

No wonder, then, that the captains of slave-trading ships who made bargains with African slave-dealers described them as 'sharp arithmeticians'.[14] According to one account, 'One of these brokers has perhaps ten slaves to sell, and for each of these he demands ten different articles. He reduces them immediately by the head into bars, coppers, ounces, according to the medium of exchange that prevails in the part of the country in which he resides, and immediately strikes the balance.' The fact that the instructions for this system of calculation were passed on by word of mouth makes it all the more impressive, but it also meant that the slave trade decimated its use. It is impossible to say how many great mathematical minds were shipped to Europe, the Caribbean and the Americas, never to be used again. In consequence, Africa's rich mathematical traditions have never been properly appreciated — except, perhaps, those that flourished in Egypt.

Fractional Gains

As book titles go, *Directions for Knowing All Dark Things* is a knockout. It sounds like a volume you might stumble across in the damp basement of an occult bookshop, a guide to the art of summoning spirits for mischievous purposes. It's not. It's an ancient Egyptian mathematics textbook.

In the West, it is better known as the Rhind Papyrus, named after the Scottish lawyer who bought it in Thebes sometime around 1858. Most of it (the full document was 18 feet long) is housed in the British Museum in London. Another part is held by the New York Historical Society. But the whole of it was written by an Egyptian scribe called Ahmos some three and a half thousand years ago. Ahmos (his name means 'moon-born') created the document as a copy of a thousand-year-old text that outlined the mathematical tricks of the Egyptian priesthood.

The ancient Egyptian kingdom ran on calculations related to the annual flooding of the Nile. Engineers would read the river's depth gauges and make reports of how the water level was rising. Astronomer-priests would keep calendars so that the citizens could prepare for the day of the helical rising of the star Sirius — the moment when its position relative to the Earth was far enough away from the Sun that it could be seen again. This was the last day of preparation for canal dredging and dyke repair.

Thanks to their calculations, the Egyptians could reliably direct the Nile's rising waters through their canals and onto the agricultural lands, where fertile silts settled on the soil. Once the waters had soaked into the land, or been directed back to the main body of the river, a new season of farming could begin — once, that is, the land had been subdivided and redistributed.

The inrushing waters washed away all boundaries and markings, so scribes had to keep records of how much land households had farmed in the previous year. Administrators then allocated them an equivalent part of the newly fertilised farmland, which they worked out using what we would see now as quite basic arithmetic. It might well have been rather basic to the ancient Egyptians, too, but it was clearly considered important enough that the scribes regularly made copies of fading documents that described the process.

Much of the Rhind Papyrus is, in essence, an introduction to handling fractions. You might be surprised to hear that fractions weren't invented to torture schoolchildren; they were an essential part of running an economy. For a civilisation that needed know how much grain was in a cylindrical store, or to carry out the government's wishes in regard to the division of land, food and wages, whole numbers — the numbers we have dealt with so far — weren't enough.

Whole numbers, or integers, or counting numbers, are how our brains match objects in our environment against an abstract concept of 'one-ness', 'two-ness' and so on, and map those to our fingers (which, if we're lucky, exist in virtual form in our brains) when we need to manipulate those quantities. Fractions are different. They are a means of dividing up the whole numbers by comparing one against another. And they are difficult: the idea of fractions of whole numbers is a terrifying leap for a brain that didn't evolve to imagine such things.

If you found fractions distressingly hard to handle in school, you're far from alone. Leonardo da Vinci, for one, would be right there beside you. For all his great accomplishments in art, invention, and astronomy, he was hopeless with fractions.[15] His notebooks reveal a man who was clumsy whenever he multiplied or divided them. He found it impossible to believe, for instance, that dividing something by a fraction of 1 (such as ⅔) would make the number bigger.[16]

Da Vinci would certainly have struggled with your education. According to the school curriculum in the US students are supposed to be able to deal with fractions by the age of 12 or 13, and should be able to, for instance, put ½, ⅝, and 2/7 in ascending order of size. Can you? Most 12- and 13-year-olds can't.

Here's another one: which of 1, 2, 19, or 21 is closest to the sum of 12/13 and ⅞? Three-quarters of 12- and 13-year-olds students in the US get that wrong.[17] The most common mistake is to add the numerators and denominators (the top and bottom numbers) separately — in other words, treat them like natural numbers. That's no surprise: it's exactly what you have been trained to do until this point. Instead, you have to estimate (12/13 and ⅞ are both close to 1, so their sum will be close to 2), or convert the fractions so that they have the same denominator and then add the adjusted numerators. When you start to think about it, you realise that fractions are cruel. We've already seen that the ability to handle natural numbers is a hard-won victory; with fractions, you then have to pull that apart.[18]

For all the difficulty we have with them, civilisation after civilisation realised that fractions were worth the effort. The Babylonians got there first, around 2000 BC, followed by the Egyptians, the Hindus, the Greeks, and the Chinese. Which means that, if I've got my maths right, a species that has existed for 300,000 years has only been using fractions for (very roughly) the last one-hundredth of its existence. If you still needed proof that there's nothing natural or innate about even basic mathematics, there it is.

For the creation of accounting, the key to civilisation as we know it, two further mathematical innovations were essential: negative numbers and the concept of zero. Despite their ubiquity today, both were hugely controversial ideas that took hundreds of years to gain the status they have now.

The Need for Negative Numbers

It is astonishing to think that we had been doing subtraction for thousands of years before anyone could answer the question 'what is 1 take away 2?' But again, we have our brains to blame. We simply can't imagine minus one apple, so there is no way for us to have an innate sense of negative numbers. They are yet another enormous leap, an imagined-from-scratch concept. But, as with fractions, they were too useful not to invent.

The historical trail for negative numbers is messy. In the *Arthasastra*, published by the ancient Indian teacher Kautilya around 300 BC, we can see that Indian accountancy was already sophisticated enough to include notions of assets, debt, revenue, expenses, and income, and there is some evidence that Indian accountants may have been representing debts using negative numbers at this time. In *The Nine Chapters on the Mathematical Art* the Chinese mathematician Jiuzhang Suanshu was calculating with negative numbers. We're not sure when it was written — the best guess is between 200 BC and AD 50 — but it describes red rods representing positive numbers and black rods representing negative numbers. However, despite this arithmetic use, Suanshu could not countenance negative numbers coming out of operations such as solving equations. They were, it seems, a purely practical device used only in trade and commerce.

In AD 628 the Indian mathematician Brahmagupta also suggested that debt could be represented by a negative number. He even presented the rules for multiplication (product) and division (quotient) when dealing with positive numbers (fortunes) and negative numbers (debts):

The product or quotient of two fortunes is one fortune.

The product or quotient of two debts is one fortune.

The product or quotient of a debt and a fortune is a debt.

The product or quotient of a fortune and a debt is a debt.

In modern terms, he is saying this:

Multiplication or division of two positive numbers gives a positive number.

Multiplication or division of two negative numbers gives a positive number.

Multiplication or division of a negative number by a positive number gives a negative number.

Multiplication or division of a positive number by a negative number gives a negative number.

You might know these rules as something like 'a minus times a minus is a plus', and so on, or 'mixed minus, pairs plus'.

Clearly, the Indian accountant was comfortable with negative numbers by this point. In the Western world, however, things progressed a lot more slowly. The problem was, the West inherited its mathematics from the Greeks, and they had been obsessed with whole numbers. They could pair them in fractions, but however small they got, they never went negative.

The first tentative exploration of negative numbers in the West was published in 1202, in a book called *Liber Abaci* (*Book of Calculation*). The author's name, Fibonacci, may be familiar to you. It wasn't his real name; it was conjured up by a biographer six centuries later. But Leonardo da Pisa was indeed the son of Guglielmo Bonacci (hence *fi* — son of — *Bonacci*), and the name has stuck so well that it now resonates as one of the great names in mathematics.

Fibonacci spent the early part of his career as an Italian customs official working in Algeria. Accompanying his father on travels to locations such as Syria and Egypt, he came into early contact with mathematics outside of the Italian tradition, discovering all kinds of operations and ideas that seemed radical, revolutionary and sometimes just plain useful. *Liber Abaci* contains a multitude of mathematical inventions, puzzles, solutions, and curiosities, including the rules (based on how fast a population of rabbits will grow unchecked) for generating the numerical series that now carries Fibonacci's name.[19] But it also contained a discussion of how you might use negative quantities as recognised mathematical tools. In his example problem, Fibonacci sets out a scenario where four men are sharing money from a purse in specific proportions:

> there are four men; the first with the purse has double the second and third, the second with the purse has triple the third and fourth; the third with the purse has quadruple the fourth and first. The fourth similarly with the purse has quintuple the first and second ...

With the four men designated *A* to *D*, and the purse as *P*, we would write this as a set of 'simultaneous equations':

$$A + P = 2(B + C)$$

$$B + P = 3(C + D)$$

$$C + P = 4(D + A)$$

$$D + P = 5(A + B)$$

These establish the numerical relationships between all the unknowns, and Fibonacci says there is a set of solutions, the smallest of which is that 'the second has 4, the third 1, the fourth 4, and the purse

11, and the debit of the first man is 1'. The interesting point, though, is in that 'debit': Fibonacci explicitly says that, 'this problem is not solvable unless it is conceded that the first man can have a debit', and goes on to demonstrate that the debit involves arithmetic with negative numbers.

Despite the success of Fibonacci's book in propagating some mathematical ideas to his fellow Europeans, it largely failed when it came to negative numbers. They didn't really catch on in the West for hundreds of years. Take French mathematician Blaise Pascal's response to the question 'what is the result of subtracting 4 from 0?' for example. To him, the answer was 0 — and he scorned anyone who thought otherwise. In his *Pensées*, he said, 'I know some who cannot understand that to take four from nothing leaves nothing.'[20] This was halfway through the 17th century; the era of microscopes, telescopes, Newton's laws, and electricity. Even in the midst of scientific discovery and technological innovation, some of the West's finest minds were reluctant to countenance the existence of negative numbers.

Things only began to change when John Wallis, Savilian Professor of Geometry at Oxford University, realised that people think better when given a visualisation. In 1685, Wallis published *A Treatise of Algebra*. Within its pages, he lays out numbers along a line, and allows those numbers to extend to a negative. He concedes that it's difficult to entertain in the abstract. But in the case of something physical like distance, he argues, you can see how it works. He didn't put it exactly like that, of course. Here are his words:[21]

> Yet is not that Supposition (of Negative Quantities) either
> Unuseful or Absurd; when rightly understood. And though, as
> to the bare Algebraick Notations, it import a Quantity less than
> nothing: Yet, when it comes to a Physical Application, it denotes
> as Real a Quantity as if the Sign were +; but to be interpreted in a
> contrary sense.

In other words, it's a positive number, but in a contrary sense. That's essentially what we would say. His 'Physical Application' involves measuring a distance along the line from a fixed point, and back again — and further. He asks how far a man would be from his starting point if he advances 5 yards from A and he goes back 8. His answer, as yours would no doubt be, is –3.

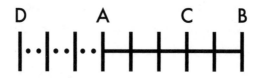

John Wallis's number line

It's fascinating to read Wallis's lengthy defence of this claim. 'That is to say, he is advanced three yards less than nothing,' he says, and goes on to explain the idea in various different ways. Where today the answer would just get a tick in a nine-year-old's exercise book, Wallis expends enormous amounts of effort exploring what he means, adding another 17 lines of text about the significance of –3. Wallis understands that the idea is a radical one.

These days we see that minus sign as one small tool in the enormous toolbox of mathematical notation. We are so used to it, and what it signifies, that we have lost our appreciation of what a vital innovation it represents. Once we had accepted their existence, negative numbers opened up more than a way to quantify debts; they provide natural, easy-to-follow mathematical descriptions for a huge variety of phenomena. Physical forces are one: dealing with positive and negative numbers allows us to predict the range of artillery shells as gravity works against their flight. The same process enables us to make strong, stable architectural structures where all the forces and loads are balanced. Wherever two things are in opposition — be it a spaceship and gravity, income and expenditure, the wind in a ship's sails and the ocean's resistance at its bows — negative numbers ease the calculations.

For all their power, negative numbers weren't quite enough to give us the modern world. You'll have noticed, perhaps, that Wallis's number line doesn't really deal with numbers: it has only subdivisions of the line marked by the letters A, B, C, and D. These correspond to what we would term 0, 5, 3 and −3, and there is good reason for Wallis's avoidance of these figures. Another hugely significant mathematical tool — zero — has yet to be accepted.

Nothing Matters

The story of zero began when King Shulgi implemented 'positional notation' in his mathematical state. We learn very early in our lives that when we write a number such as 1,234 we can assign different values to the individual digits depending on their position. The 4 is the lowest position, and designates four units, equivalent to something like four apples. Because we work in what mathematicians call base 10 — that is, we group our numbers in tens — the next digit is the tens, and there are three of them, making thirty. Moving to the left, we have the result of ten of the previous column: ten tens, or a hundred. There are two of these in 1,234. Finally, we have one group of ten hundreds: which is one thousand. Hence we read the number as one thousand two hundred and thirty-four.

King Shulgi's positional notation was in base 60, not base 10. It's not clear why this became such a dominant technique in the early days of number-writing. Some historians of mathematics argue that it comes from the fact that 60 is a number that can be divided into whole numbers by all the integers between 1 and 6 (and six other numbers). That certainly makes it easy to work with, especially when dividing up goods, costs, and measures. Others suggest that it might be to do with it making things easy when dealing with the approximate number of days in the year. Whatever the reason, it has left a legacy: the Middle Eastern kingdoms that eventually became Babylonia are responsible for our 360 degrees in a circle, 60 minutes in a degree, 60 minutes in an hour and 60 seconds in a minute.

The Babylonian base-60 notation is similar to our base-10 system: 34 is written as three 10 marks and four unit marks, for example. But the notation marks only go up to 59, so the base-10 number 424,000 would be written as 1574640 in base 60. That's 40 units, 46 lots of sixty, 57 lots of sixty sixties (60^2) and 1 lot of sixty times sixty times sixty (60^3).

Using this notation (and ours) is all very well as long as there is nowhere in the number that has *no* 'lots' of something. But, returning to base 10, what about the number 4,005, where there are no hundreds and no tens? We had to find a way to denote 'none of these' when we wrote it down. And so began the use of what we now know as a zero.

It wasn't always a zero. There is a lot we don't know about this story, but for the Babylonians, this placeholder for an empty column seems to have started out life as a slanted wedge shape: ◤ (though even this is contested).[22] The Maya and Inca also used a placeholder, an abstract symbol or glyph, to denote an empty column. Neither of these is the zero we are familiar with: that symbol came to us — we think — through the Hindu *shunya*, a dot that represented nothingness. The oldest known use of this round placeholder for an empty column is the Bakhshali manuscript, an Indian text written on 70 leaves of birch bark. It has been dated to somewhere between AD 224 and 383, and might have served as a training manual for Buddhist monks. The *shunya* still took a while to become a mathematical zero, though. Brahmagupta's AD 628 treatise that embraces negative numbers is also the place where zero — in this case, the Hindu *shunya* — is first seen as more than just a space. Here, it is part of the number line, a quantity in its own right that is bound by the same laws of arithmetic that govern the others. Here is Brahmagupta's take on how zero interacts with other numbers, both positive and negative:

A debt minus zero is a debt.

A fortune minus zero is a fortune.

Zero minus zero is a zero.

A debt subtracted from zero is a fortune.

A fortune subtracted from zero is a debt.

The product of zero multiplied by a debt or fortune is zero.

The product of zero multiplied by zero is zero.

It was the 10th-century Persian mathematician and astronomer Muḥammad ibn Musa al-Khwārizmī who first brought zero to the attention of the West. In his books he employed what we now know as the Hindu-Arabic numerals, and included zero, celebrating its usefulness in keeping positional notations on track. He designated it *sifr*, or 'empty'. In Latin this became *zephyrum*, from which the Italians derived the name zero.

Zero was not only a notation tool for al-Khwārizmī, however. Like Brahmagupta, he also used it as a tool in algebra, cementing its importance in the manipulation of numbers, and called it 'the tenth figure in the shape of a circle'. For al-Khwārizmī, zero was clearly one of the numbers, and it plays a key role in his AD 830 volume *Al-kitab al-mukhtasar fi hasib al-jabr wa'l-muqabala* (*The Compendious Book on Calculation by Completion and Balancing*). It is from this title that we get the term 'algebra', by the way, and we get the word 'algorithm' from the author's name; al-Khwārizmī was a man of genuine influence. He saw his book as something that everyone could use, providing numerical tools useful 'in cases of inheritance, legacies, partition, lawsuits, and trade ... where the measuring of lands, the digging of canals, geometrical computation, and other objects of various sorts and kinds are concerned'. However, despite this vast landscape of applications, Western minds baulked at the idea of zero.

Zero seems so obvious and familiar to us now that it is hard to imagine numerical systems working without it. But they did — and for a very long time. When the French monk Gerbert of Aurillac travelled to Spain to study Islamic mathematics in the 10th century AD, he saw the zero, and ignored it. Gerbert saw the value of al-Khwārizmī's mathematical ideas, and went on to disseminate many of them to European traders. But he didn't bring zero to Europe, choosing instead to focus on teaching abacus skills.

Even two hundred years after Gerbert's journey, zero was still unwelcome: the English historian William of Malmesbury is said to have referred to the idea as 'dangerous Saracen magic'.[23] And even when Fibonacci introduced the European mind to the power of zero, he stopped short of including it among the numbers. In his *Liber Abaci*, Fibonacci tells us that 'The nine Indian figures are: 9 8 7 6 5 4 3 2 1. With these nine figures, and with the sign 0 ... any number may be written.' Calling zero a 'sign' shows that he, unlike al-Khwārizmī, had still not truly incorporated it into the numbers.

It's hard to say why, exactly. Partly, the resistance is due to a sense that an absence of something can't be treated in the same manner as a presence. Just as Greek mathematical philosophy saw no place for negative numbers among the sacred nature of the positive whole numbers, it also had no tolerance for the idea of making nothing into an entity worthy of attention. Aristotle pointed out in his text *Physica* that you couldn't meaningfully divide by zero, and so it couldn't count as a number.[24] Perhaps even more importantly, though, zero had no place on the abacus, the primary calculation tool of the intelligentsia of medieval Europe.

Abaci weren't always beads or stones threaded on wires, as we conceive of them now. The name is thought to derive from ancient Middle Eastern words for dust and board; it seems they may have first involved covering a surface in dust that could be marked with a finger or with stones, then cleared to begin a new calculation.

The abacus avoids the need for a zero by dint of its arrangement. Because you can see the stones or marks laid out in a neat row, you implicitly gain positional information without needing to explicitly mark a 'nothing here'. And once you had learned the full set of algorithms for operating the abacus, you would certainly be resistant to a newfangled way of representing numbers.

It used to be the case that a knack for abacus work was a much sought-after skill. It was even a little bit sexy. When Geoffrey Chaucer wrote 'The Miller's Tale' as part of *The Canterbury Tales*, he made every effort to make his protagonist a shameless (in every way) intellectual. 'Clever Nicholas' has an astrolabe for making astronomical measurements and a Greek-language textbook to guide his thoughts. And, Chaucer tells us, he keeps the stones of his abacus neatly arranged on the shelves by his bed: he is always ready to perform any necessary calculations. Effectively, he is a nerd. The fact that he succeeds in cuckolding his landlord, a rich but dull-witted carpenter, would be a surprise twist in today's culture. But in 'The Miller's Tale', Chaucer sets Nicholas up as irresistible to the carpenter's beautiful young wife.

Scholars suggest that Nicholas was everything Chaucer's close friend King Richard II admired. At the time he wrote *The Canterbury Tales*, Chaucer was a close associate of the king and, more interesting to our purpose, the chief auditor of the customs for the port of London. The abacus is there for a reason: its possession marked out the intelligentsia of the 1380s — Chaucer included.

These days, abaci are manufactured in various forms: the Chinese *suanpan*, the Japanese *soroban*, and the Russian *schoty*, for instance. They are still used in many of these places to teach young schoolchildren how to visualise basic arithmetic processes, and there is even evidence that the abacus can reshape the brains of those who use it.[25] The best abacists of today — mostly East Asian schoolchildren — have used them so much that many of them have no need for the physical thing. They can imagine the positions and movements of the beads in much the same way that a seasoned chess player can play out a game without a board or pieces. Experienced abacists can do much more than add and subtract; they can use their instrument to find square roots of numbers, for example. But, for all the wonder of the abacus, we haven't needed one for many centuries — largely because zero exposed their limitations. Writing down your mathematics, complete with as many zeros as you need, frees you to work with numbers of unlimited size and calculations of unlimited complexity.

The earliest official use of zero and Hindu-Arabic numerals in the West seems to have been in the 1305 accounts of the Gallerani firm in Pisa, Italy.[26] Roman numerals, however, remained in fashion, and dominated accounting for the next century — traders and bankers are resistant to change. But gradually, people realised that Roman numerals, and other number systems without zero, make heavy work of arithmetic. When the Hindu-Arabic number system came along, verifiable, written calculations became possible. With numbers written out using the numbers from 1 to 9, and complemented by 0, we could develop algorithms — recipes for calculation — that would make light work of multiplication and division of huge numbers. The abacus gradually became redundant, and by 1500, the administrators of the prestigious Medici bank had adopted a clear policy: only Hindu-Arabic figures were to be used in their accounts.[27] Slowly, inexorably, their influence spread. Within a few hundred years, the Hindu-Arabic figures, including the ultimately irresistible zero, had taken over the world.

It's no accident that this coincides with an unprecedented acceleration in the development of human society. With zero and negative numbers in our toolbox we found ourselves able to keep track of the numbers that ushered in an era of global trade and prosperity — exemplified by the Medici Bank, the French Revolution, and the glorious financial innovations of Alexander Hamilton.

Balancing the Books

The acceleration, you'll be surprised to hear, began with double-entry bookkeeping. At its simplest, this is a way of ensuring that no errors have crept into the accounting. Each transaction is recorded in two separate accounts, making it possible to check them against each other. The essentials are laid out clearly in Luca Pacioli's 1494 book *Summa*, which we mentioned in the discussion of finger signs for numbers:[28] 'All the creditors must appear in the ledger at the right-hand side, and all the debtors at the left. All entries made in the ledger have to be double entries — that is, if you make one creditor, you must make someone debtor.'

The earliest use of the system was probably in the records of Korean merchants. According to documents still kept by the Taehan Ch'on'il Bank, they used something called the *sagae Song-do chibubop*, or 'four-sided Kaesong ledger' method in 11th-century trades with China and Arabia. The four sides are the receiver's name, the giver's name, the commodity or cash received, and the commodity or cash disbursed. Every transaction required a double entry.

Sadly, there's no direct evidence of this; the Taehan Ch'on'il Bank's records are little more than anecdotes, and the earliest surviving Korean mercantile records are from the mid-19th century. But we do have evidence of a 15th-century text describing double-entry bookkeeping. The Croatian mathematician Benko Kotruljević who was born in Dubrovnik in 1416, wrote *On Merchantry and the Perfect Merchant* in 1458.[29] Kotruljic lays out a system where every transaction has two mentions in the ledger. If you bought a measuring stick, you would enter the value of the stick as a credit in one column, and the amount you paid for it as a debit in another.

Europe was already using the system before Kotruljic's book was published, though. We know this because we have various examples, including some of the financial records of a Venetian merchant called Jachomo Badoer.[30] Documenting the period from 1436 to 1439, Badoer's double-entry ledgers (or something close to them) are written entirely in the Hindu-Arabic script, complete with zeros, and detail his transactions in Constantinople. There, he exported spices, wool, slaves and various other commodities to Venice, where his brother ran the import and sales side of the business.

Badoer's business was just one of hundreds, if not thousands, operating in the financial powerhouse created in the northern region of 15th-century Italy. It was here that the trade routes between East and West met, where European Crusaders stopped on their way to and from Jerusalem, and where myriad currencies had to be negotiated in commercial transactions. Implementing a system that could keep track of all the numbers involved — including the concept of debt carried by negative numbers — was always going to give businesses an edge.

As well as facilitating trade, double-entry bookkeeping changed the way businesses grew and operated. The double-entry system is centred on the accounting equation *Assets = Liabilities + Owner's Equity*. In other words, the health of the business is the sum of its debts and its current holdings, which are all accounted after every transaction. This allows anyone associated with a business to know at a glance what it is worth. So if you were considering whether to loan money or goods to a company, you could see exactly what you were getting involved with in terms of debts, operating costs, assets, loans, and net worth. No more taking the owner's word for it, or trusting the family name. If the books balance, and you like what you see, you can go ahead with your transaction. That extends to buying a business too. Because double-entry bookkeeping is based on the premise that a business is an entity that is distinct from its owner, the owner can value and sell their business if they so choose. It is hard to comprehend just how revolutionary this would have been when the system first came into widespread use. There was no longer any need to keep things in the family and stick to one line of business; you could build a firm and treat it as you would disposable capital. If you fancied starting another business alongside your first, the books could prove your business acumen, and even be used as security against a loan.

Accurate bookkeeping spawned other industries too, such as maritime insurance. Crossing seas and oceans was a risky business, with pirates and acts of God just waiting to steal or sink valuable cargoes. The ability to keep accurate accounts of onboard assets made it easier to assess risks and underwrite voyages. The ships themselves also became more valuable; systems of accounting for ownership and tax liability made it possible to protect them from seizure by monarchs, who were often desperate for assets that would fill up emptying war chests. The merchant classes agreed to provide royal households with all due taxes and to do the necessary paperwork. All they asked in return was an agreement that anything recorded as a private asset could not be commandeered by a monarch. Onshore, the same thing happened with agricultural land: the ability to designate, prove, and transfer ownership created a market for the land and the labour that would render it profitable — as well as freedom from arbitrary redistribution of land by the ruling classes.

As the centuries passed, double-entry bookkeeping's open, accessible numbers facilitated the rise of capitalism. John D. Rockefeller, billionaire owner of Standard Oil, was among its biggest advocates: he began his career as a bookkeeper, and frequently attributed his success to his strict control and deep understanding of his balance sheets.[31] Rockefeller once confessed that he gained much of his business acumen from poring over his first employer's ledgers from the years before he worked there. Bookkeeping was such a joy to him that he made an annual celebration of 26 September, the date he got his first position as an assistant bookkeeper in 1855.

The pottery magnate Josiah Wedgwood provides another example of the power of good bookkeeping. In 1772 he carried out a deep and complex analysis of his struggling firm's double-entry accounts and used them to turn his business around — to spectacular effect.[32] The numbers in the ledgers enabled Wedgwood to identify where spiralling costs and late payments were crippling his burgeoning enterprise. He implemented various measures — including the first foray into mass production —

that would maximise his profits. Though he became extraordinarily wealthy, the money Wedgwood made did not merely line his own pockets. He channelled it into various social concerns, most famously the campaign for the abolition of slavery. Another great legacy of the Wedgwood fortune came when it enabled Josiah's grandson by marriage, a budding naturalist, to voyage around the world aboard the HMS *Beagle*. The control over numbers — and hence profit — conferred by double-entry bookkeeping funded the development of Charles Darwin's theory of evolution by natural selection. Who could ever have predicted that the human invention of numbers would lead us to a deep understanding of human origins?

Finally, it's worth noting that Karl Marx was also captivated by bookkeeping. While researching the origins of capitalism, Marx asked his friend Friedrich Engels, whose family owned a cotton mill in the north of England, to supply him with 'an example of Italian bookkeeping with explanations'.[33] It was through his studies of the Engels firm's accounts that Marx came to share Rockefeller's view that the control and optimisation of production costs, through the understanding that came from controlling the books, was a pivotal part of the capitalist enterprise. Marx, it's safe to say, was not a fan. He saw capitalism as a means by which a few individuals — the owners of the means of production — could amass wealth. And that wealth came at a cost: it could develop 'only by simultaneously undermining the original sources of all wealth — the soil and the worker', Marx said. Marx laid the blame for capitalism's power squarely at the feet of double-entry bookkeeping. He who controls the numbers, he realised, controls everything and everyone.

Marx's mention of how capitalism would undermine the ground beneath our feet was strangely prescient. In recent years, commentators have begun to suggest that the environmental crisis — catastrophic climate change, unprecedented extinction rates, accelerating deforestation, and severely reduced soil fertility (among other concerns) — also has its roots in double-entry bookkeeping. Our obsession with the power of numbers has led us to value only what we can write down as figures on a spreadsheet. This has caused us to both devalue any assets that we cannot represent in manipulable numbers, and to create measures that are not up to the task for which they were created. We thus reduce the economy of nations to a single, arbitrarily defined number — the gross domestic product — which, according to orthodox economists at least, must be maximised at all costs. Meanwhile, we operate the global economy through the institutions ruled by numbers: the banks, which operate with near-impunity as they control the fortunes of entire nation states. But at the same time we fail to account for the value of our soils, our forests, our wildlife — especially the insects — and transnational assets such as the polar ice-caps. It is, some have suggested, an algorithm for environmental collapse.[34]

Not that corporations are a bad thing, and banks are certainly not bad in and of themselves. On the contrary: without banking services, many of us would not have homes, or be able to live our lives of comparative luxury. It is no wonder that so few of us have even thought about opting for a life outside of the capitalist system. But there is a downside. Numbers and their accessories may make balance sheets possible. Balance sheets may make auditing and accountability possible. But in the absence of honest auditing, they also make it possible to conjure up imaginary transactions, to fantasise about moves that might increase profits, and to find ways to make those moves even when the money involved doesn't exist. The problem is, sometimes that fantasy money gets called upon — and then the banks fail.

We did not learn our lesson from the collapse of the Medici Bank; the last four centuries have seen the influence and importance of those who own our numbers grow ever stronger. In the 2007 financial crisis, each of the failing banks was thought to have assets — money, property, and liabilities, essentially — so huge that just a few of them added together were worth more than the countries in which they were trading. The economic wisdom was that, if governments had let them topple and fall, these banks would have crushed too many people's livelihoods, too many of the economy's most vital companies, and too many of their host nation's hopes of economic growth. And so they were deemed 'too big to fail', and had to be bailed out at enormous cost, plunging the entire world into chaos.

That chaos is testimony to the power that numbers have gained over us since we invented them. It explains why an accountant could provoke the French Revolution, why the American struggle for independence was shaped around banking practices, and why Europe was plunged into financial turmoil on the death of a single medieval book-keeper.

But for all the achievements and influence of accounting, many of us would measure the development of civilisation by more glamorous indices. Architecture, painting, sculpture, and music, for instance, are often hailed as the hallmarks of sophistication. But here too we have to acknowledge a strange coincidence. Both accounting and art experienced a world-changing relaunch at the same time and in the same place: northern Italy, at the start of the Renaissance. As we have seen, the rise of accountancy can be traced to developments in the art of counting. But it was another mathematical innovation that gave us the better-known glories of the Renaissance. To explore this revolution, we must return to ancient Greece, the birthplace of a subject that seemed utterly banal to me in my first years of mathematics education. What on Earth, I used to wonder, is the point of geometry? We are about to find out.

Chapter 2

GEOMETRY

How we conquered and created

It started out as a search for the perfect forms and numbers on which the universe had been constructed, but geometry didn't remain a study of abstract shapes for long. Once you grasp the nature of the triangle, for instance, you can map the heavens and the Earth, and navigate your way to unfathomable riches. Add in the mystical properties of the circle, and you can build, paint, or conquer anything that takes your fancy. The story of geometry starts with superstition, then evolves through an era of reckless greed and ambition, before delivering our greatest works of art — and a means to carry the world in your pocket.

Have you ever heard of Prester John? Perhaps you caught Shakespeare's reference to him in *Much Ado About Nothing*, where Benedick alludes to the elusiveness of this strange figure. 'Will your Grace command me any service to the world's end?' he says. 'I will go on the slightest errand now to the Antipodes that you can devise to send me on; I will fetch you a toothpicker from the farthest inch of Asia; bring you the length of Prester John's foot; fetch you a hair off the great Cham's beard; do you any embassage to the Pygmies, rather than hold three words' conference with this harpy.'

No one ever saw Prester John, let alone his foot. He was long thought to be a king reigning somewhere in Africa, an Arthur for the medieval era.[1] Although no one knew where he lived — or if he still lived, since the legends were told for hundreds of years — kings, popes and emperors sent him messages and entreaties to join forces with them to fight the Muslim threat. That's because they believed that Prester John's kingdom ran with rivers of emeralds, was littered with gold, and was home to good Christian men who could fight as well as any warrior who ever lived. Prester John, in short, was the Christian ruler that all the other Christian rulers wanted as their best friend. Unfortunately, that was almost certainly impossible, since they had fallen victim to what seems to have been a centuries-old hoax.

In 1165, the Byzantine Emperor Manuel Comnenus forwarded a letter to the Holy Roman Emperor Frederick Barbarossa. The letter purported to be from a 'Prester John' — Prester being a title akin to 'priest' — who told of his kingdom in 'India'. He was, the letter said, a fabulously wealthy descendant of one of the three magi who had visited the infant Christ. Excited, Frederick wrote back immediately, sending the letter in the care of an emissary. We don't know what became of this missive, but it was only the start of the disappointments associated with Prester John.

By the time the 15th century rolled around, Prester John fever was running high. In 1400, for instance, King Henry IV of England wrote a letter to 'Prester John King of Abbysinia', seeking to open Anglo-Abyssinian relations. (In case you're wondering, the passage of more than two centuries didn't matter, because Prester John's original letter had mentioned that he owned fountains of eternal life.) In 1402, a Florentine man called Antonio Bartoli presented himself at the palace of the Doge of Venice, capturing everyone's attention. Bartoli had brought with him several Africans, plus some pearls, leopards, animal skins, and exotic herbs. He claimed to be the envoy of Prester John, lord of India, who wanted to strike up an association with the Christian leaders of Europe. The Doge sent him back with gifts including a silver chalice, a fragment of the Holy Cross and 1,000 ducats, as well as several skilled craftsmen and weapon-makers.

Presumably, Bartoli made off with the loot, as he was never seen or heard of again. Nonetheless, word of Prester John's riches, and his desire to see Christians push back the Muslim insurgence, had spread far and wide. The legend eventually reached Portugal, finding fertile soil in Prince Henry, the devout, ascetic, intellectual third son of the king. Almost immediately, Henry decided he would be the one to finally locate Prester John and recruit him to the Catholic cause, even if it meant mapping and navigating the whole Earth.

There is a great deal of scholarly controversy over the exact nature of Henry's effort.[2] Some say that it was an established school at Sagres, complete with academic staff and lectures for mariners, navigators, and ship designers. Others contend that it was a looser, less formal affair. Either way, Henry's goal was to harness all the mathematical knowledge of southern Europe with the aim of conquering the oceans and discovering Prester John's elusive kingdom. Henry brought a slew of experts to Portugal to train a whole generation of naval men in the sciences of shipbuilding, navigation and map-making. He did well enough that the Italian scholar and papal secretary Poggio Bracciolini praised the Prince's achievements: 'How illustrious indeed to have been the only one of such courage, of such resolution and of such planned purposefulness to have dared to do that which none so far had undertaken or attempted,' he wrote in 1448. 'You alone [have discovered] unknown seas in regions never seen, unknown races living outside the known world and savage peoples living at its farthest confines beyond the regular annual shifts of the sun's track, where none before had opened a way.'

You won't be surprised to learn that Henry never actually found Prester John. But he did set everything in place for Europeans to conquer the world. How? By harnessing the geometry we all learned in school: the sines, cosines and tangents of right-angled triangles, and the relationships between the circumference and diameter of circles and spheres.

In Henry's hands, geometry became a means of mapping, navigating, and dominating the world. After being shipwrecked off the coast of Portugal, Christopher Columbus made his home there, for instance, and made full use of the maps, training, and scholarship available because of Henry's Sagres programme. You already know what came of that. In the centuries that followed, the same knowledge gave us something better still: the golden ages of art and architecture.

The Hidden Power of Triangles

As an eight-year-old, I was obsessed by the legend of King Arthur. I remember stumbling my way through T.H. White's *The Once and Future King*, and imagining which of his knights I most wanted to be (my preference changed almost daily). I can also still picture the room where, at the same age, I was first made to study geometry.

The word 'geometry' literally translates as the 'measurement of the Earth'. But in school it tends to be limited to the study of two- and three-dimensional shapes and their properties. That mostly works out to triangles and circles, but you can build other things too, like squares, cones, pyramids, and, if you're feeling adventurous, dodecahedrons. Then you might look at graphs, how to cut lines and angles in half, and techniques for measuring distances between points on a line. My overwhelming feeling about all this was the opposite of obsession. I found geometry dull. I was willing to admit there was something slightly intriguing about Pythagoras' theorem. But once I had learned this, that the square of the hypotenuse — the longest side of a right-angled triangle — was equal to the sum of the squares of the other two sides, I switched off and let my imagination rejoin King Arthur at his

Round Table, where all are equal. Now *there* was an inspirational use of geometrical shapes.

Perhaps my eight-year-old self would have been more interested if my teacher had introduced Pythagoras as a figure of legend, something like a Greek King Arthur. It turns out there is no solid evidence that a man ever lived who did all that Pythagoras is supposed to have done. He has been variously credited with influencing the rise of vegetarianism; realising that the morning and evening stars were both Venus; recognising that the Earth is a sphere; and suggesting that the planets move according to mathematical equations. However, we know almost nothing for certain about Pythagoras because none of his writings have survived.[3] We don't even know simple facts such as his birthplace — we *think* he was born on the Aegean island of Samos, the son of a gem-cutter. The only thing that scholars are fairly confident about is that someone eventually established a collective of scholars in his name on the island of Croton, in modern-day Calabria.

This was a school captivated by numbers. The members of this sect, who were bound together by secret oaths, entered the school through an arch that declared 'All Is Number'. As far as the Pythagoreans were concerned, numbers ruled the cosmos. The level of this obsession is illustrated by a story — possibly also a legend — about the fate of a scholar who broke his oath to the community.

The tale begins, like all good geometric tales, with a right-angled triangle. Two of its sides have length 1. According to Pythagoras' theorem, we square each of the two shorter sides (*A* and *B*), add those squares together, and we have the square of the hypotenuse *C*. We can write this as an equation:

$$A^2 + B^2 = C^2$$

Since *A* and *B* are both 1, that means:

$$1 + 1 = C^2$$

If C^2 is 2, then C is the number that can be multiplied by itself to give 2: the square root of 2, which is written as $\sqrt{2}$.

To us, that's no big deal. But to the Pythagoreans, it was a huge problem. The only way they could write down any number was if it could be expressed as what mathematicians call an integer — a whole number like 1, 2, 3 and so on — or as the ratio of two integers. You'll have used mathematical ratios: we bake using ratios of flour to fat, for example, and we make cocktails using ingredients in fixed ratios. In a Manhattan, say, we use two measures of bourbon to one measure of sweet vermouth: that's a 2 to 1 ratio, which can also be expressed as a fraction: ½ as much vermouth as bourbon. The Pythagoreans tried to find a ratio of two numbers — expressed as a fraction, such as ⅓ or ⅚ — that would be the numerical equivalent of $\sqrt{2}$. But, try as they might, they couldn't.

Then things got worse. Someone in the Pythagorean School managed to prove that their frantic search for a solution to the square root of 2 problem had been futile. It turns out that $\sqrt{2}$ simply *cannot* be expressed as the ratio of two integers, no matter how hard you look for them. That's because it is what we now call an 'irrational' number — a number that can't be written as a ratio. Pi is another, and there are many more, some of which play significant roles in modern mathematics.

The Pythagoreans were so appalled at this insult to the universality of whole numbers that they agreed to keep the existence of irrational numbers a closely guarded secret. But according to the legend, a Pythagorean called Hippasus told someone outside their sacred circle. When Hippasus's crime was discovered, he was thrown off a boat and left to drown in the middle of the Adriatic Sea. The moral of this story is clear: triangles, at least the right-angled ones, are a matter of life and death.

While the activities of Pythagoras and his sect are cloaked in uncertainty, there is someone concerned with triangles who definitely did exist: Thales of Miletus. Thales was a highly intelligent and opportunistic man who lived on what is now the western coast of Turkey. He was born around 640 BC, and is now known as the father of scientific philosophy. To his contemporaries, though, he was a famously shrewd businessman. One story has it that, in a growing season that looked set to produce a bumper olive crop, Thales had the foresight to corner the market in olive presses. He rented his presses out to growers at exorbitant rates. If they refused to rent from him, he bought their unpressed olives at a knock-down price. Thales eventually amassed a fortune that allowed him to retire in middle age. He then devoted the rest of his life to academic study. Philosophy, science, and mathematics were the beneficiaries of Thales' newfound leisure time: he pioneered the use of testable hypotheses and theories in explaining the natural world, and was the first to write down several of the central propositions of geometry that we still study today.

Thales almost certainly learned these propositions during his travels in Egypt. Thousands of years previously, geometers had been central to Egypt's great construction projects, such as the pyramids. But Thales was able to add to the Egyptians' geometry by giving practical demonstrations of various principles. He showed, for instance, that what is now called the 'isosceles' triangle has two identical angles at its base. He did this by demonstrating that an exact copy of the triangle, flipped onto its back, still remained a perfect copy. He also showed that everything you need to know about a triangle can be determined from knowing its base length and the two angles of the sides to the base. This is useful information. If you want to know how far out to sea a ship is, you can construct a triangle with the ship at its top corner. Take a known length of the shoreline as your triangle's base, and stand at one end of this base and measure the angle between the base and the ship. Then walk to the other end of the base, and again measure the angle to the ship. Now create a smaller triangle — inscribe it in the sand of the beach, if you like — that has the same angles between the base and the top corner. Calculate its height to base ratio, then multiply that by the distance you measured along the shoreline for the original triangle. Now you have the ship's distance offshore.

Similar triangles can tell you the distance to a boat offshore

Using a variant on this technique, Thales showed that there are useful proportions to be exploited in such 'similar triangles'. He reportedly wowed the Egyptian King Amasis II by calculating the height of a pyramid from the height of a stick placed at the end of the pyramid's shadow.

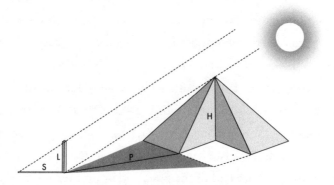

Thales showed how to find the height of a pyramid by measuring shadow lengths

The length of the pyramid's shadow P and the length of the stick's shadow S are in the same ratio as the height of the pyramid H and the length of the stick L. We can write it like this:

$$\frac{P}{S} = \frac{H}{L}$$

Then we can rearrange the formula to give:

$$H = \frac{PL}{S}$$

So the height of the pyramid is given by the length of the stick multiplied by the length of the pyramid's shadow divided by the length of the stick's shadow. We shall see that this kind of calculation eventually formed a central pillar of medieval navigation known as the rule of three: if you know three things about your similar triangles, you can calculate a fourth, unknown thing, and the world is at your feet.

Thales' favourite discovery with triangles involved making one within a circle. He showed that if you take the diameter of a circle (the width at its widest point) and use it as the base of a triangle with a corner at the circumference (the perimeter that defines the circle), that corner is always a right angle. According to legend, Thales was so shocked by this insight that he sacrificed an ox to the gods in gratitude for the revelation.

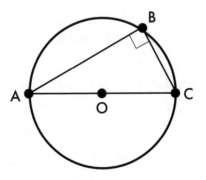

A triangle inscribed inside a circle has a right angle at the circumference

Thales' method of inscribing a triangle inside a semicircle to create a right angle was undoubtedly put to use in the construction industry. It would involve choosing a line for your building — perhaps the perfect north–south line that you get by noting where the shadow of a tall pillar such as an Egyptian obelisk falls at noon. Then you tie a rope to a peg, and knock the peg into the ground at the point on that line where you want your corner (A). Now use the other end to score a circle in the Earth. Where that circle meets the north–south line at B, you score another circle of the same radius. Where that intersects the first circle (C), draw a long line from B, through C, and as far again to D. Join A, B and D together and you have a right-angled triangle with AD running perfectly east to west. What better way to start building a temple to the Sun?

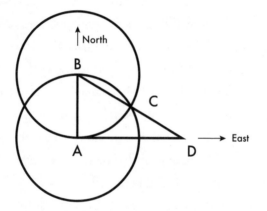

A geometric method for finding east–west from north–south

Though basic enough for eight-year-old children to learn, Thales' method wouldn't have been the first that builders employed to create a right angle. Records show that, somewhere around 2000 BC, scholars in what is now Iraq were using the theorem that we mistakenly attribute to Pythagoras. The fact that a strict mathematical relationship exists between the side lengths of a right-angled triangle was a cornerstone of the construction industry. To create a building that has perfectly square corners at its base, construction workers grabbed a rope and some pegs.

They would divide the rope up into 12 units, then peg one end into the ground. Run three units of the rope to where you want the wall with a square corner, and put a second peg in. Now turn through roughly 90 degrees and run four units of rope. Knock in the third peg here, then bring the rope back to the first peg. If it doesn't quite meet up, move the third peg until it does. Then you'll have a perfect right angle at the second peg because $3^2 + 4^2 = 5^2$.

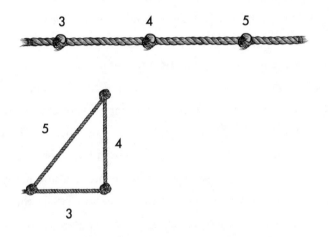

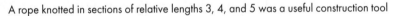

A rope knotted in sections of relative lengths 3, 4, and 5 was a useful construction tool

It was understanding these properties of triangles, circles and angles that gave us our first glimpse of the size of our planet. In 240 BC, Eratosthenes, the chief librarian at Alexandria in Egypt, used this mental toolkit to calculate the circumference of the Earth.

According to ancient accounts (these, it should be said, were written centuries after our hero's death), Eratosthenes heard that on one day of the year, the noonday sun lit the entire shaft of a deep well in the southern Egyptian city of Syrene (now Aswan). That day was the summer solstice, when the sun is at its northernmost extreme, and thus directly over cities that lie on the line of latitude we now call the Tropic of Cancer. Eratosthenes reasoned that this information, combined with a measurement in Alexandria, would allow him to calculate what proportion of the Earth's circumference lay on the (roughly) north–south line between Alexandria and Syrene. On midsummer's day, he set a stick vertically in the ground using a plumb-line. At noon, he measured the angle that the stick's shadow made with the vertical. It was 7.2°. Since a sphere covers 360°, Eratosthenes knew the surface distance between Syrene and Alexandria must be 7/360 of the whole circumference. He knew this distance to be 5,000 stadia, and used the rule of three to calculate the circumference. His answer would have been around 250,000 stadia.

I would love to be able to tell you how accurate Eratosthenes's result was. Unfortunately, we don't know exactly how these stadia convert to modern measurements so we can't be sure. But it certainly lies in the right ball park. The modern value for the equatorial circumference is roughly 24,000 miles. Eratosthenes's measurement is probably somewhere between 24,000 and 29,000 miles. That's a significant achievement for a man whose nickname was 'beta', or 'second-best', because, though brilliant at many things, he was never actually number one at anything.

And Mr Second Best didn't stop there. He understood that the axis on which the Earth spins to give day and night is not quite parallel to the axis of the orbit around the sun. This is why we have seasons: the angle of the tilt means that at certain points in the Earth's journey around the Sun, the northern hemisphere is receiving more intense sunlight than it will be receiving six months later. Eratosthenes used the geometry of shadows to calculate what this tilt might be, and got a value of 11/83 × 180°; that is, 23.85°. The actual value is around 23.4°. Again, not bad.

Sine, Cosine, and Tangent

We can't talk for long about triangles before we have to face the unholy trinity: sine, cosine, and tangent. Very few of us have a clear understanding of these three words. Put simply, they are numbers related to the lengths of sides of a right-angled triangle. These days, we encounter them most frequently as buttons on a calculator. It wasn't a whole lifetime ago that they were more commonly to be found as tables of numbers printed in booklets: my first geometry teacher handed one of these booklets — I remember its red-and-white covers — to each student at the beginning of the lesson. I also remember seeing sines, cosines, and tangents as nothing more than a means of finding the answer to a pointless maths problem.

What do they even mean? It's not clear when these exact terms came into common use, but it's likely that variants of the values they represent have been around for many thousands of years. Remember Ahmos the Egyptian scribe? His copy of the Rhind Papyrus includes a question that goes like this: 'If a pyramid is 250 cubits high and the side of its base 360 cubits long, what is its seked?' From the solution he gives, involving ratios of the side lengths in right-angled triangles, we know that a seked is what we call a cotangent — the inverse of the tangent. In this case it's the inverse of the tangent of the angle between the base of the pyramid and its face. But we're getting ahead of ourselves; let's start with the sine.

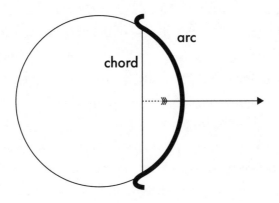

How the sine got its name

The name comes from a mistake. It started out as a description of the straight vertical line in the illustration above. It's called a 'chord' of the arc; the arc is that portion of the circumference that looks like the bow of a bow and arrow. The Sanskrit word for the chord is the same as for a bowstring: *jiya*. Arabic translations rendered this as *jayb*, but were written down, following tradition, as just *jb*. Translators putting the ancient geometry texts into Latin mistook this for *jaib*, meaning bosom or breast, and used their word for this: *sinus*.

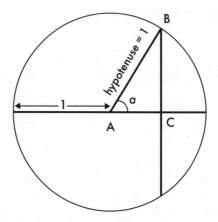

Where sines, cosines, and tangents come from

But what is it, exactly? Sines, like cosines and tangents, are just ratios — comparisons, if you like — of the lengths of a triangle's sides. The sine of the angle *a* is the ratio between the length of the vertical side of the triangle (BC) and the radius of the circle (which is the hypotenuse of the triangle, AB in the illustration). In other words, the sine of the angle is the length of the opposite side divided by the length of the hypotenuse. The complementary sine — cosine — of the angle *a* is the ratio between the base of the triangle AC and the radius AB. The tangent of *a* is the ratio of the length of the vertical side of the triangle (BC) to the length of its base (AC). And with this knowledge safely on board, we are ready to sail out of sight of land.

Finding the Way

'Navigation is nothing more than a right triangle,' said the French mariner Guillaume Denys in 1683.[4] He is referring to what we call a right-angled triangle: the properties of this shape, he says, are all a sailor needs to understand. It was a truth that had been established for hundreds of years, starting with the windrose lines of Mediterranean sailors. Windroses were lines on a chart that connected ports or other places of interest. When you drew the chart correctly, the angle of the line, relative to north, gave you a compass bearing to follow.

Sailors collected these into *portolani* — port books — which were widely used to navigate the Mediterranean in the 13th century. However, it was rare to be able to sail from port to port in a straight line. And when ships went off course during a journey, whether it was because of unfavourable winds, an island in the way, or because they were waylaid by pirates, the crew would use the mathematics of triangles — trigonometry — to set them back on course. That's why they carried tables of sines and cosines, or an instrument that could calculate them, such as the sinecal quadrant.

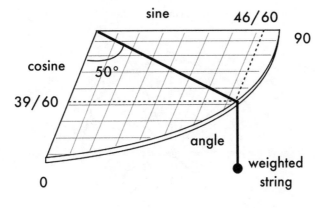

A sinecal quadrant

Our first description of the sinecal quadrant comes from the 9th century AD. That's when al-Khwārizmī, head librarian at the House of

Wisdom in Baghdad (and the man who introduced us to zero in the previous chapter), laid a square grid onto a quarter circle, with a string pinned at the origin. The other end of the string reaches to the curved edge, which is divided into 90 degrees. The two straight edges are marked with a scale divided into 60 units; one is used for calculating the sine of the angle, and the other for calculating the cosine.

These days, you can download a printout of the sinecal quadrant from the internet. It is astonishingly easy to use: essentially, you set the string to the angle, and follow the line up to the sine or cosine edge to get your reading. But if you were a sailor who didn't want to engage with sines and cosines at all, you could just use the *toleta de marteloio* — a trigonometry table designed specifically for maritime use. This told you how to correct your course if the wind, or something else, had diverted your journey. You just put in how many miles you've sailed off course, and how far off your desired heading you've been sailing. The *toleta* then gives the distance to travel on your new heading before you're back on track.

This relies on another chart: the compass rose. It has each quarter of the rose divided into eight 'rhumbs', which describe direction. The first quarter, for instance, has north-by-east, north-northeast, north by northeast, plain northeast, and so on.

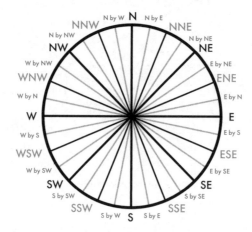

The windrose chart. Each of these named directions is a 'rhumb'

Let's see if we can think like a 13th-century sailor. Imagine you are wanting to sail from Athens to Heraklion in Crete. It's approximately 212 miles south-southeast, but the wind will only allow you to sail due south. As you move through the Aegean Sea, you would work out the distance you have travelled by 'dead reckoning', the art of estimating speed by watching waves pass a ship, or by throwing some wood overboard and counting the time it takes for the wood to pass a known length of the hull. After you have sailed 75 miles, the wind shifts; now you can sail east-southeast. But for how long should you sail on that bearing to return to your original intended course?

As Denys said, it's just right-angled triangles. And with a *toleta de marteloio*, you don't even need to worry about trigonometry. If you know the number of rhumbs away from your original intended course your actual course was, plus how far you sailed, the *toleta* gives you the distance off course. Then you choose the appropriate number of rhumbs between your original intended course and the 'return' course you are about to sail, and you have an indication of how far you should go along the return. Finally, the *toleta* gives you the distance left to sail along that original intended course once you reach the endpoint of your return course.

Toleta de Marteloio

Difference of rhumbs	Off course	Course made good	Return course	Course made good
	For every 100 miles sailed	For every 100 miles sailed	For every 10 miles off course	For every 10 miles off course
1	20	98	51	50
2	38	92	26	24
3	55	83	18	15
4	71	71	14	10
5	83	53	12	6 1/2
6	92	38	11	4
7	98	20	10 1/5	2
8	100	0	10	0

A simple *toleta de marteloio*, which helps sailors correct their course

We have sailed 75 miles due south, which is 2 rhumbs off the ideal course. Using the *toleta*, we work out that we are therefore $^{75}/_{100} \times 38$ miles off course — that is, 28.5 miles.[5] We will be sailing our return (to the original course) at 4 rhumbs off the original, ideal course. How far should we sail on this heading? The answer is $28.5 \div 10 \times 14$ miles — that is, 40 miles. This will get us back to the point where we can use the original, ideal heading to travel the remaining distance to Heraklion.

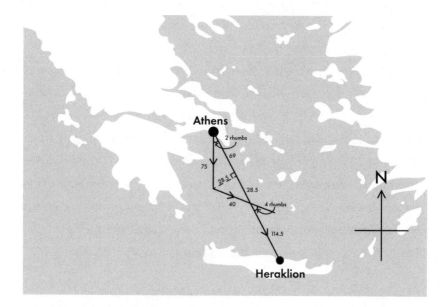

Navigating a route from Athens to Heraklion using rhumbs and a *toleta de marteloio*

So, assuming all is now well, we will have sailed 75 miles due south, then 40 miles east-southeast, and from there we must sail 114.5 miles south-southeast. If we can't, we just rinse and repeat. We can check our progress on a chart, looking out for any shallow areas ahead where we might run aground, and be sure that we won't sail endlessly until we run out of food and water.

Making Maps

These triangular tricks and tables were such an essential part of a mariner's toolkit that they became quite a money-spinner for the educational entrepreneur, who would set up a school for sailors or produce a textbook. The truly savvy teachers would do both — requiring every student to purchase a copy of their book. French mathematician Guillame Denys was able to monetise his knowledge of trigonometry so adroitly that he set up a sailing school in Dieppe. There he educated French Navy trainees, private sailors, and even pirates. Denys' Royal School of Hydrography was just one of many such European institutions in the 16th and 17th centuries. Sailors might have a reputation for ignorance, illiteracy and brutish behaviour, but many of them were accomplished mathematicians nonetheless.

They had to know their limits, though. The flat triangle geometry that worked for Mediterranean *portolani* doesn't work the same way for longer journeys. Because the Earth is (roughly) a sphere, its surface is curved, and its triangles are different. To see how this works, score three straight lines into the skin of an orange to make a triangle, then peel it away. It doesn't look quite triangular, does it? The sides of the triangle have a kind of bulge, and if you were to add up the three angles at its corners, you would find that they add up to more than the 180° of a triangle drawn on a flat plane. One result of this is that sailing across an ocean along a constant compass bearing doesn't take you along anything approaching a straight line on the Earth's surface. It actually takes you along what is known as a loxodrome: a spiral that winds its way around the globe, always crossing the north–south meridian lines at a fixed angle.

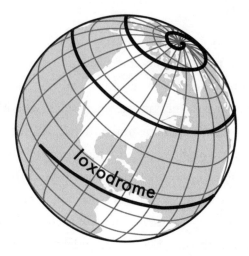

Follow a fixed compass bearing and you'll spiral around the globe on a loxodrome

This means that even if I know the compass bearing of Bristol, England from New York, USA, sailing the journey in that direction will not be the fastest possible route. Instead I have to take the shortest path between two points on a sphere: the circumference that passes through both points and has the centre of its circle at the centre of the globe. Applied to navigating the surface of the Earth, it's known as a 'great circle'.

Now imagine we are planning a great circle journey between New York and Bristol, and trying to provision our ship.[6] In order to work out the distance we must sail, we have to imagine a spherical triangle that has one of its corners at New York; another at Bristol, and the third at the north pole. If we know the latitude of New York and Bristol (that is, how far above or below the Equator they sit), we can work out the distance between them using standard trigonometry. However, it's a long and tedious process. It involves imagining a host of triangles, some of which stick out from the centre of the Earth to beyond its surface. We would have to perform a series of complicated trigonometry calculations on these triangles, each one of which might go wrong and land us in trouble. The alternative was to go to sailing school, where the lecturers would teach you the relevant shortcuts.

The complication introduced by the spherical nature of the Earth's surface is the central problem of map-making. Geometers have long understood that the Earth's surface features can't be directly translated onto a flat surface such as a map without some kind of distortion occurring. For many thousands of years, map-makers have sought 'projections' of the spherical surface that minimise just how misleading their map might be. A projection takes the latitude and longitude and performs a mathematical operation on them so that, when drawn on a flat plane, the angles and distances between all the various points make sense. The maths involved is a combination of the geometry of spheres and trigonometry (and in the modern day, calculus, which we'll come to in a couple of chapters' time).

The first world map projection we know of is credited to Agathodaemon of Alexandria, who lived (we think) in the 2nd century AD. Ptolemy, a Greek mathematician who lived in Alexandria about the same time, published this map in his book *Geographia*. This is a projection with latitude and longitude lines, which was revolutionary at the time — but Agathodaemon's lines of latitude are curved, and his longitudes are not parallel, but splay out from the northernmost point.

In roughly the same era, a cartographer called Marinus of Tyre produced the 'equirectangular projection' on local maps. In this projection, latitude lines are made horizontal and longitude lines are vertical on the flat sheet, and all the lines set at an equal spacing. This, with a few tweaks here and there, was enough to keep navigators happy for more than 1,000 years.

Christopher Columbus was a map-maker too, and had been charged with making exquisitely accurate maps on his journeys. By the end of the 15th century, the Spanish and Portuguese royal courts understood that there was vast wealth to be gained for anyone who could repeatedly make safe passage to the East Indies, or across to the Americas. That meant creating geometrical constructions that would allow map-makers to issue a precise set of instructions for navigators to follow. Columbus's

1492 diary, addressed to his sponsors, shows his intention to produce something definitive:[7]

> I have it in mind to draw a new nautical chart on which I will locate the sea and the lands of the Ocean Sea in their proper places, each beneath its wind, and moreover to write a book and put the likeness of everything in it in drawings, in their equinoctial latitudes and their longitudes from the west. It is above all important that I forget about sleep and pay great attention to navigation, for in this way duty is fulfilled; and these things will be hard work.

However, the fundamentals of spherical geometry mean that no flat, planar map of a sphere is perfect in every respect. Take the world map that is probably the most familiar to you, for instance: the projection produced by Gerardus Mercator. Created in 1569, it became so dominant because it was a boon for sailors. Mercator's handling of spherical trigonometry kept the angle between any two points on his map entirely as it is on the spherical surface of the Earth, meaning that compass bearings on the map translated to compass bearings for the ship. There was a downside: land masses — and thus distances — far from the Equator were grossly enlarged. The world does not actually look like it does on a Mercator projection; Alaska, for instance, is one-fifth the size of Brazil, but Mercator made them look the same. He also made Greenland look the same size as Africa, even though Africa is 14 times larger. But if you're just sailing oceans not too far north or south, who cares?

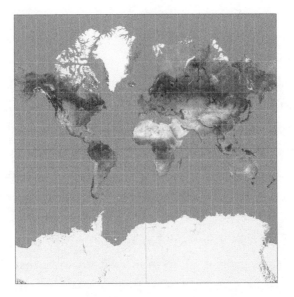

The Mercator projection
Strebe, CC BY-SA 3.0, via Wikimedia Commons

Maps are very different now, of course. These days they are 'dynamic': you can change their properties to suit your needs, as you do with the GPS map on your phone. Their usefulness lies in the fact that their compromises can change depending on the navigator's requirements. This is no mean feat in mathematical terms: the mathematical tricks required to achieve this eluded NASA's best scientists for decades.

The man who eventually worked it out was the unsung hero John Parr Snyder. Perhaps you had already heard of Ptolemy and Mercator, but I'd be very surprised if you had heard of Snyder. It's a shame: according to *The New York Times* he was 'every bit the equal of any map maker in history, including Gerard Mercator'.[8] It would certainly be hard to find anyone whose head for geometry has had a more direct impact on your life.

In the best possible sense, Snyder was a nerd. In 1942, aged just 16, he began to keep notebooks that detailed interesting things he had discovered about geography, astronomy, and mathematics.[9] Among many other things, the notebooks contained facts about triangles, and his thoughts and insights about the geometry of flat surfaces and solid objects. These thoughts quickly led to a fascination with map projections. Snyder was captivated by the way mathematical equations could transform points on the globe to points on a flat plane, and by what the maths did to the geometrical relations between those points. But he never studied the subject formally. At university, he studied chemical engineering, and that was the profession he entered. It was only decades later, in the 1970s, that he entered the sphere of professional map-making.

In 1972, NASA launched *Landsat-1*, the first satellite built to study the Earth's geography. To insiders, it was clear that the satellite could also provide an entirely new kind of world map, and two years later the cartographic coordinator for the United States Geological Survey (USGS) published a paper describing a suitable mathematical projection. Alden Colvocoresses — Colvo to his friends — imagined a map that would account for the movement of the satellite's scanner, the satellite's orbit, the rotation of the Earth, and the way the axis of that rotation evolves in a 26,000-year cycle thanks to Earth's 'precession'. In order to avoid distortions, the map would have the form of a cylinder, and the surface of this cylinder would oscillate back and forth along the cylinder's long axis. In this way, there would be no disastrous distortions as the data from the satellite was compiled into a map. It was an audacious idea. But no one at NASA or the USGS knew how to do the geometric analysis required to actually construct the projection.

Snyder first heard about the problem in 1976, after his wife bought him a rather nerdy 50th birthday present: a ticket to attend 'The Changing World of Geodetic Science', a mapping convention in Columbus, Ohio. Colvo gave the keynote, and outlined his problem. Snyder was hooked. He spent five months of his evenings and weekends solving it, using his spare bedroom as a study and nothing more technical than a Texas Instruments TI-56 programmable pocket calculator. Almost immediately, the USGS gave Snyder a job.

Snyder's projection is called the space oblique Mercator projection. According to one expert, it is 'one of the most complex projections ever devised'. Among other operations, it involves applying 82 equations to each of the data points. The result is an output that creates a Mercator projection, but from a moving vantage point, and with only minimal distortion of the area directly below the satellite. We can't even begin to contemplate how it works here, but it is intriguing to note that Snyder's paper laying out the ideas behind it involves a complex array of sines, cosines, and tangents. Thousands of years after we first discovered its properties, we are still harnessing the power of the triangle.

The space oblique Mercator projection was an essential step towards constructing satellite maps of our planet. These are vital for everything in 21st-century civilisation, from military operations and navigation to weather forecasting, environmental conservation, and climate monitoring. Snyder's projection gave us Google Maps, Apple Maps, the satnav in your car, and every other digital mapping technology you can think of. Finally, we had God's own view of the Earth. This achievement had taken 600 years — assuming you consider that it started with Prince Henry the Navigator's urgent search for Prester John.

Pi and the Circle

Obsessed as I now am with triangles, it would be remiss of me to only touch tangentially on the properties of circles. These too have played a vital role in our story.

As with triangles, human interest in circles has always had a very practical motivation. Calculating the areas of triangles and rectangles told ancient rulers how much to tax a landowner, because any field, whatever its shape, can be split roughly into rectangles and triangles. That makes it easy to calculate the field's total area, and that gives the taxman the information on taxes due. In the same way, calculating the volume of cylindrical pots or grain silos (or even just conical piles of spices) is also relevant to raising levies on goods grown, purchased, or manufactured. And calculating those volumes involves understanding circles.

The first practical concern is getting an accurate enough value for the ratio of the circle's circumference to its diameter. This is the ratio known to us by the Greek letter π (pi); the circumference of a circle is π times the circle's diameter. Many ancient cultures didn't worry too much about accuracy. The Babylonians and earliest Chinese geometers went with 3.0, and the Egyptians were using 3.16 around 1500 BC. Archimedes found his approximation of it by first drawing a multi-sided shape (a polygon) inside a circle. He then split that shape up into triangles whose base is one of the polygon's sides (the other two sides being radii of the circle). By calculating the area of each of these 'isosceles' triangles, you can work out the area covered by the polygon. The more triangles you draw, the closer the polygon comes to covering the area of the circle, and the more accurate your value of π. The area of all those triangles is roughly equivalent to the area of the circle, which is πr^2. Since you know the radius r, this gives you a value for π. By 240 BC, when Archimedes was looking at the properties of wheels, he had pegged π at somewhere between 3.140 and 3.142 (by using a 96-sided polygon).

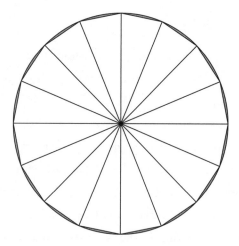

Archimedes' method for calculating π using triangles within a circle

Around 450 AD the Chinese geometer Zu Chongzhi created a 24,576-sided polygon to calculate π as lying between 3.1415926 and 3.1415927. These days we know π to be 3.14159265358979..., and have calculated its value to trillions of decimal places.

If aliens were to land on Earth, they might be surprised by just how fascinated we are by π. No other number has been so obsessively studied. It is the subject of feature films and documentaries, music, and art. Perhaps I'm too biased towards triangles now, but I'm struggling to understand where π's appeal comes from. Perhaps because its digits never end, and have no discernible pattern? That's reasonably beguiling, since a circle has no end either — but is it *that* different from the never-ending digits of the square root of 2, which came from a triangle?

Admittedly, the usefulness of π cannot be questioned. It pops up literally everywhere: in maths, physics, finance, architecture, art, music, and engineering, to name just a few areas. That is because it is intimately involved in any mathematical description of a phenomenon that repeats. If you want to do the mathematics of waves — whether they arise in sound, water, applications of electromagnetism, stock market data, or any other medium — you're effectively looking at something whose properties go around in circles, and so you'll need π. But since we are looking for the influences of mathematics on our civilisation, let's not ignore one of the most overlooked applications of π: in architecture.

Construction by Numbers

If you have ever visited the Hagia Sophia in Istanbul in modern Turkey, you were probably too overwhelmed by its beauty to notice the mathematics of the building. But it's there, because the structure was designed by two mathematicians: Isidore of Miletus and Anthemius of Tralles.[10]

The Emperor Justinian came to Isidore and Anthemius because he wanted a new, stone-built church on the site of a church that had just been destroyed by rioters. Tens of thousands of people had been killed while protesting, among other things, high taxes (what else?). Justinian was after something imposing, something that would stamp his authority on the city.

The Hagia Sophia

Public domain, via Library of Congress Prints and Photographs Division, Washington, DC 20540

The design for the Hagia Sophia was perfect for the job. Its geometry is an impressive, complex mix of an 82-metre-long rectangular basilica combined with a central square structure that is dominated by a 56-metre-high dome. Nothing so ambitious had ever been attempted. Completed in AD 537, it was the largest building on the planet at the time. It became the pride of Constantinople and was quickly acknowledged as one of the architectural wonders of the world. How did the architects accomplish this astonishing feat in just six years of construction? By using approximations of π and the square root of 2 — and by following geometrical shortcuts created by Heron of Alexandria.

Heron was born around AD 10, and became a celebrated mathematician and inventor. He came up with ways to pump water, find the area of a triangle, make a steam engine, and many other achievements. His architectural digest, *Stereometrica*, would have been well known to Isidore and Anthemius, who taught in the universities of Alexandria and Constantinople. *Stereometrica* is a practical book that tells the reader how to work out the volumes and surfaces of various architectural constructions and thus budget and plan for the necessary materials and their shipping. It also tells the reader how to fudge their way through a project.

The most illustrious architectural structures require curves, circles, and spherical geometry. And this means using π. But, as we have seen, π is one of those irrational numbers that just didn't exist for the Greeks. They couldn't write it down, and they certainly couldn't pass it on to a stonemason. No wonder Heron wrote that 'the numbers are uncomfortable for measuring' before suggesting an approximation for π. He suggested $^{22}\!/_7$, then used examples that set the radius or diameter as a multiple of 7, making it easy to cancel the denominator (lower part) of the fraction when dealing with the various attributes of a vault or dome. Heron was a master of using geometry to make an architect's life easier. A 'pendentative' dome, like the one in the Hagia Sophia, is constructed in two parts. There is the hemisphere at the top, supported by a slightly larger vault that is made up of curved triangular sections. Heron explains the exact calculation method for such spherical triangles: subtract four spherical sections from a hemisphere after inscribing half of a cube within the hemisphere. That makes it easy to calculate volume (and thus weight), and surface area (and thus the amount of plaster that will be required). What's more, by the end of this section, Heron has reduced the number-work to the point where he offers 'standardised solutions' for the construction industry, minimising the maths a site foreman would have to do in his head.

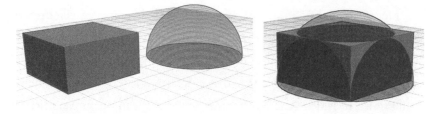

The dome of the Hagia Sophia is derived from a half-cube cut from a hemisphere

It was originally thought that the dome of the Hagia Sophia was built on a square that had a side length of 100 Byzantine feet — a lovely round number. We don't know exactly how big a Byzantine foot was, but modern measurements set the side length at 31m, which certainly puts the definition of a Byzantine foot within the acceptable variation if the square was 100 feet along each side. However, if you have a square whose side is a pleasing 100 feet long, the length of its diagonal — the distance from one corner of the square to the opposite corner — will reflect the fact that half of this square constitutes a scaled-up version of the triangle that caused the Pythagoreans so much grief. In other words, you'll have to deal with the fact that the diagonal length is a multiple of the square root of 2. That might end up giving the diameter for a vaulted dome sitting atop the square, like that at the Hagia Sophia, as 141.421356237... feet. No Byzantine surveyors' instruments would have been able to cope with this irrational number.

Let's think like Heron and start with the circle calculations. If we want to make them easier, it would make much more sense to set the diagonal, and thus the diameter of the dome, at 140 feet. 140 is a multiple of 7, and thus compatible with the $^{22}\!/_{7}$ approximation of π.

If the diagonal was going to be 140 feet, Isidor and Anthemius would have needed to tell the foreman what the length of the square's sides needed to be in order for him to begin the process of construction. They probably did this using a Pythagorean trick called the side-and-diagonal number progression, which gives an approximation for the factor $\sqrt{2}$.

You start with a square of side 1, and call its diagonal 1. Obviously, this is a *very* rough approximation. To improve on it, you create a bigger square in order to approximate that square's diagonal. The side length for the next square in the series comes from adding the previous side and diagonal together (so, 2 in this case). To get the next diagonal in the series, add the previous diagonal to twice the previous side (giving 3).

Arrange this next diagonal over the next side as a fraction, and you have ³⁄₂, which is 1.5 — a slightly better approximation to $\sqrt{2}$. As these squares get bigger, you get ⁷⁄₅, ¹⁷⁄₁₂, ⁴¹⁄₂₉, ⁹⁹⁄₇₀, and so on ... which means they get progressively closer to a precise value for $\sqrt{2}$. For example, ⁹⁹⁄₇₀ is 1.41428... which is not bad at all (remember, $\sqrt{2}$ is 1.41421...).

Isidor and Anthemius would then have used whichever of these fractions they — and their foreman — could best cope with. That gave *S*, the side length of the square for which they knew the diagonal length. It's a similar triangle to the one of sides 1, 1 and $\sqrt{2}$, so the ratio 1 to $\sqrt{2}$ is the same as the ratio of *S* to 140. Substitute in ⁹⁹⁄₇₀ for $\sqrt{2}$, and

$$\frac{S}{140} = \frac{1}{99/70}$$

Using the rule of three, you can solve this to show that *S* is a squeak under 99 feet. In the circumstances, 99 feet is a perfectly good approximation for the side length, and the builders of the Hagia Sophia will have no difficulty marking out that square. And because the diagonal of the square below the dome was a multiple of π, the dome would have been relatively easy to construct.

It is even possible that Isidor and Anthemius didn't have to do this much maths. It's likely that Heron created tables where you look up the diameter of your dome, and find all the relevant numbers for constructing its constituent elements. None have survived, but we have books where he created similar tables for other purposes. What's more, he drew diagrams to help the architect — diagrams that look suspiciously similar to the dome and some of the vaults in the Hagia Sophia.

Isidor and Anthemius almost certainly used such shortcuts. We have several texts that make reference to a (sadly lost) commentary that Isidor wrote on Heron's calculations for designing and building vaulted structures. And let's be clear, Heron was standing proudly on the shoulders of other mathematicians, particularly Archimedes; there are plenty of other examples around the world of ancient geometric ratios being put to work. Durham Cathedral in the northeast of England was clearly built using approximations of the ratio of the square's side to its diagonal. The committee that designed Milan Cathedral in the late 14th century enlisted the help of a mathematician, Gabriele Stornoloco, to discuss whether the construction should be *ad quadratic* or *ad triangulum* — that is, should be built according to the ratios of the diagonal to the square, or according to the ratio of the altitude to the side length of the equilateral triangle.[11] Stornoloco went for triangles — supplemented by squares, rectangles, and hexagons. Academics argue about how he calculated the necessary ratio of an equilateral triangle's height to its side, which is $\sqrt{3}/2:1$. As with the Hagia Sophia, no stonemason was going to work out the necessary values, but it seems that Stornoloco gave the tradesmen just three specific dimensions — the breadth of the nave and four aisles; the height of the triangle indicating the apex of the nave-vault; the interaxial distance between the nave-piers — and the ratio 26/30. Other medieval European structures, notably the cathedrals at Reims, Prague, and Nuremberg, invoked the side-to-diagonal ratio of the regular pentagon, which involves using known approximations of $(\sqrt{5} + 1)/2$.[12] This was just how things were done; why reinvent the wheel? After all, once the basic calculations are in the bag, the rest is as easy as π.

A Ray of Light

The Hagia Sophia is a designated wonder of the ancient world, one of many architectural marvels that were built in antiquity. So why is it that all the paintings that are generally considered the world's best only began to be created a thousand years later, starting in the 15th and 16th centuries? And why did this revolution in painting coincide with the conquest of the oceans and European mapping of the globe? Is it a coincidence? No: they both took advantage of newly rediscovered mathematical arts that had been lost during the centuries of holy war.

At the beginning of the 7th century, Islamic nations began to spread into and conquer much of western Asia, and northern Africa. By the end of the century, they were even making headway into Europe, settling in Spain and the Balkans. In the 11th century, though, they pushed Christendom to breaking point. Now Christians were banned from visiting Jerusalem, the holy city. In 1095, Pope Urban II responded by instigating the first of the crusades. There were seven more crusades over the next 200 years, and they were far from successful. The Muslims remained very much in control of Jerusalem and all surrounding lands. It was this dire situation that made the hope-kindling stories of Prester John so powerful. But it led to more than the great trigonometry-powered navigations: it also brought us the golden age of art.

In the 1260s, an English Franciscan friar called Roger Bacon wrote a call to arms, with the aim of raising Christendom to its feet.[13] Christians would, he suggested, take back Jerusalem through a better knowledge of geometry. One application of the art, he said, would be to resurrect the fabled 'burning mirrors' of antiquity. Legend has it that Aristotle was able to use huge concave mirrors to focus the Sun's rays onto enemy ships, causing them to ignite; crusaders should be able to do the same, Bacon said. He also suggested that geometry would serve to reawaken sleeping Christians' zeal through art: surely, images created using the inherent beauty of the Lord's geometry could not fail to stir passions? 'I count nothing more fitting for a man diligent in the study of God's wisdom than the exhibition of geometrical forms,' Bacon wrote in his *Major Work*. This section was called 'On the value of optical marvels in converting the infidel'.

One theory among scholars is that Bacon was suggesting a revival of the ancient art of theatre set decoration. This would allow inspiring religious plays to rouse the fighting men of Europe against the Saracen threat. Bacon suggests 'the Latins' had skills that should be emulated; perhaps he meant Vitruvius, a Roman architect who lived in the 1st century BC, and was aware of the skills of theatrical backdrop painters:

> If a fixed centre is taken for the outward glance of the eyes and the
> projection of the radii, we must follow these lines in accordance
> with a natural law, such that from an uncertain object, uncertain
> images may give the appearance of buildings in the scenery of the
> stage, and how what is figured on vertical and plane surfaces can
> seem to recede in one part and project in another.

Vitruvius is talking about what we would call perspective. Our word derives from a Latin word meaning 'to see through', which is why we could also call it optics: the study of how light travels through, or is reflected or refracted (bent) by various media. In the ancient and medieval worlds, the words 'perspective' and 'optics' were interchangeable.

The story of optics and perspective goes back to a giant among geometers: Euclid. Around 300 BC, this Greek scholar wrote the seminal textbook on mathematics. It was called *Elements*, and remained the bestselling text — apart from the Bible — for more than 1,000 years. Only slightly less popular was the book he called *Optics*. In this, Euclid describes how light travels between objects or scenes and the eye, perhaps traversing lenses or being reflected in mirrors along the way. Many of Euclid's observations will be familiar to you, almost as common sense. He suggests that light travels in straight lines, for instance, and that you'll see an object that is higher because of light travelling along a higher path.

Euclid believed that the rays of light come from the eye, rather than the object being seen. It wasn't an uncommon view at the time, and was entirely consistent with the geometry of his theory of vision. His rays formed a light cone emanating from the eye; only objects within this cone could be seen. The 'visual rays' spread out as they move away from the eye, and so become less dense, making things that are far away look blurry.

It all worked well enough for the time, and in follow-up texts, Euclid managed to explain a whole host of phenomena, such as reflection by flat, concave, and convex mirrors, and how lenses create optical effects such as magnification. Thanks to his ability to reduce optical phenomena to issues of lines, triangles, and arcs, Euclid could apply all his geometric knowledge to create what seemed an entirely adequate theory of visual perception.

Then Ptolemy stepped up. Sometime around AD 165, he took Euclid's works and tweaked them. The major change was the idea of a line, rather than a cone, that emitted from the eye. He did some geometrical work with triangles and circles, and managed to reduce some of Euclid's errors (and create some new ones) in the calculations of where a reflected image appears — in front of or behind a curved mirror, for instance. For the next thousand years, Euclid and Ptolemy's geometric take on light rays had optics sewn up.

Yes, a thousand years. It is perhaps hard for us to understand how things could progress so slowly, but the harsh truth is that there were few ways to exploit the knowledge of optics. People had been making basic mirrors and lenses since antiquity, but they were not of high enough quality to be useful as, say, reading aids. Things only began to change on this front as the Christians recruited geometry and optics for their cause.

Not that Christians were the only ones interested in geometry and its uses. During the time of their surging conquests, Muslim scholars rediscovered Euclid, and translated his texts, adding their own commentaries. The scholar Ibn al-Haytham produced a particularly influential commentary he called *The Book of Optics*, a seven-volume textbook written between 1011 and 1021. Here, al-Haytham lays out ideas such as the visual rays forming a triangle composed of smaller similar triangles. Geometric projections then explain why objects become smaller as the visual ray comes closer to the eye, allowing large things to be seen through the eye's tiny pupil.

Al-Haytham's triangular 'visual rays' allow large things to be seen through a small pupil

The Latin translation of al-Haytham's book, entitled *De perspectiva*, was hugely influential once it reached Europe. But, despite Bacon's optical call to arms and the hordes of craftsmen who were able to make ever-improving mirrors and lenses, Europe did not see a radical shift in military success. Instead, it witnessed a revolution in art.

Getting Some Perspective

You could break your bookshelves with the volumes that have been written about the birth of linear perspective, so we will be sketching just a little of geometry's influence. One moment stands out as a useful starting point: the day Filippo Brunelleschi positioned himself 5 feet 9 inches inside the central door of Florence's Santa Maria del Fiore cathedral.

Despite being able to pinpoint the location to the inch, we don't know exactly when this was; all we can say is that it was probably in 1425. At this point in his career, Brunelleschi is a renowned architect, and designing the cathedral's dome. His vantage point inside the cathedral's door looks out on the Florentine Baptistery, which lies across the street. The Baptistery is an octagonal building with clear geometric lines highlighted in its decorations. And, according to his biographer Antonio di Tuccio Manetti, Brunelleschi paints it with perfect perspective.[14] So perfect that, when he is finished, he smugly allows viewers to compare his painted panel, which is approximately 12 inches square, against a view of the real Baptistery reflected in a mirror held beside the painting.

The Florentine Baptistery
Christopher Kaetz, Public domain, via Wikimedia Commons

Brunelleschi does this by drilling a small hole in the panel. 'The hole was as tiny as a lentil bean on the painted side, and it widened conically like a woman's straw hat to about the circumference of a ducat or a bit more on the reverse side,' Manetti tells us. Brunelleschi then has the viewer look through the hole, with the painting facing away from them, while holding up a flat mirror at arm's length. They see the reflection of the painting. Then Brunelleschi tells them to lower the mirror. Through the drilled hole they now see the real Baptistery. There was, Manetti tells us, very little difference.

Manetti is writing, of course, long after the revolution in perspective painting has been established, and no one knows whether his report is coloured by later developments in technique and technology. But Brunelleschi almost certainly used the mirror as the basis for his painting. It provided a two-dimensional projection of the three-dimensional Baptistery, avoiding the need to work out the best way to represent the way in which the human eye interprets the light coming from the building. In the 1460s, Antonio Averlino wrote about Brunelleschi that 'it was certainly a subtle and beautiful thing to discover by rule from what the mirror shows you'.[15] Averlino (who used the pen-name Filarete) goes on to give step-by-step instructions on using a mirror to draw with proper linear perspective: 'Look in it and you will see the outlines of the thing more easily, whatever is closer to you, and that which is farther away will appear to diminish.'

All the rules that we learn as children when we study drawing — near figures being larger than ones that are further away; parallel lines converging in the far distance — can actually be learned by tracing a mirror reflection. And once that had been appreciated, the mirror wasn't strictly necessary. Now it's just geometry — the kind of geometry that Euclid worked out. If you're painting, say, a temple scene, you decide on the viewer's sighting point, and draw the geometric arrangement of visual rays from the features in the subject to the viewer's eye. Then you place a flat plane — a canvas, say — at the point where the picture will be. Look where the visual rays from the temple's features intersect that canvas: that's where you draw those features onto it. In 1435, Leon Battista Alberti laid out all the steps necessary to achieve perfect linear perspective in a book that he dedicated to Brunelleschi.[16]

Clearly people were still following the recipe in the following century; Albrecht Dürer's 1525 woodcut *Man Drawing a Lute* shows the same process going on, with fine threads being used as stand-ins for the visual rays. This was the only way to achieve a realistic 'foreshortening' of curved surfaces such as the body of a lute.[17]

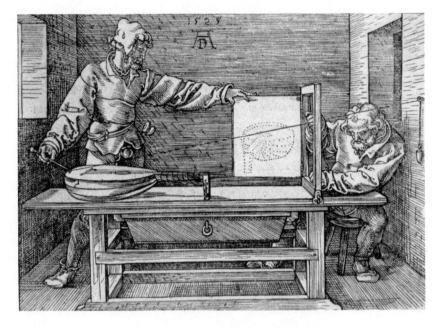

Man Drawing a Lute by Albrecht Dürer
Albrecht Dürer, Public domain, via Wikimedia Commons

Unless, that is, you used a *camera obscura*. The geometry of this instrument was first described by one of the architects of the Hagia Sophia: Anthemius of Tralles. In AD 555, he constructed a light ray diagram that showed the path followed when light rays are reflected from a mirror into a small aperture. But it was al-Haytham who described the full *camera obscura*. In his *Book of Optics*, he explains what happens when a candle faces 'a window that opens into a dark recess, and when there is a white wall or (other white) opaque body in the dark recess facing that window'. Essentially, the wall displays an image of the candle, an image that a gifted artist could easily turn into a permanent painting by putting a canvas on the wall and working on it *in situ*.

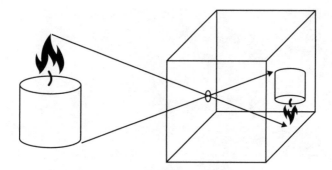

Al-Haytham's *camera obscura*

Linear perspective painting, whether by geometric sketches or by optical instruments, was a revolutionary technology. Inventors could use it to create realistic-looking representations of their contraptions, allowing craftsmen to make accurately gauged parts for the device (and, in some cases, to point out that the device wouldn't work before anyone had even lifted a tool from their workbench). You only have to look at da Vinci's sketchbooks to see the value of a perspective drawing. But its greatest immediate impact was in the world of art: in the Renaissance period, numerous painters began to create pictures that used the new rules to produce remarkably lifelike representations of their subjects. We don't know how much of it came from mathematically constructed lines drawn onto the canvas or panel, and how much of it was traced onto projections from lenses and mirrors. The artist David Hockney has argued that this uncertainty is not surprising: these were trade secrets. After all, people earned a living from their ability to emulate Brunelleschi's methods, so it made as much sense for a painter to divulge the tricks of the trade as it would for a modern-day magician to explain theirs. Some, though, were open to persuasion. In 1506, for instance, Albrecht Dürer wrote to a man called Pirkheimer that he was planning a trip to Bologna, where he would 'learn the secrets of the art of perspective, which a man there is willing to teach me'.[18] Dürer presumably paid handsomely for the lesson.

Few people would pay for lessons in perspective these days — it's all understood, and you can find Brunelleschi's insights in any number of books. For the most part, thanks to computer-aided design (CAD) software, we can also build structures without actually doing trigonometry, spherical or otherwise. To my knowledge, the only people — apart from mathematicians — in our world who still use geometry are those tasked with building new worlds. Visual effects designers creating computer graphics for Hollywood films, for instance, might still get their protractors out occasionally. Programmers developing realistic physics for videogames still find themselves pressing the sine button on their calculators from time to time. But for the rest of us, geometry is history.

In fact, the main value of geometry today may be in developing the kinds of connections between your brain's neurons that allow you to think in the abstract. The ability to hold a half-cube in your head, and wrap a hemisphere around it to create a mental picture of the dome of the Hagia Sophia, say, is perhaps not useful in and of itself. But it might be exactly the kind of skill that enables you to solve an entirely unrelated problem. I'm embarrassed to say, I can't do it, though. Or I couldn't, until I used some CAD software to create the half-cube and the hemisphere, and put them together in the same virtual space. Then I could not only see the parts of the hemisphere that Heron was subtracting, I could also spin the whole thing around, look over and under the resulting construction, and — finally — *get* it. So, for me, one question remains: how did Heron and Euclid *get* geometry?

The ancient geometers had far fewer resources for visualising their constructions than we do now. But somehow their brains were wired for doing geometry in a way that mine certainly isn't. I could say the same about Eratosthenes. I try to imagine myself outside of the Earth, looking at the positions of the Sun and the pole star, the Earth's axis and the way the globe spins through day and night. I should be able to see how the shadows change through the day, giving a way to measure the Earth's circumference. I should be able to visualise a way to measure the tilt of the axis. And that should also enable me to see how the movement of the Earth's surface relative to the stars provides a way to work out exactly where on the surface I am.

Most of us can't really do any of this — at least not without enormous effort. Why? I suspect it's because we live in a 21st-century technological society where all of these geometrical phenomena have been embedded into software. Heron, Euclid, and Eratosthenes had no little models and no visualisation software. They had no choice but to train their minds to imagine the complexities of geometry if they wanted to reap the benefits it offered. Their achievements are a testimony to the power of the human brain — something we might well forget in the modern age where so much is done for us. I can open a CAD program if I want to design a geometrical structure. I can open an app on my phone if I want to know where on Earth I am, or how to get to a specific point. Airline pilots are still trained in working out rhumb lines and loxodromes in case their GPS fails, but the rest of us have no need of the ancient ways. I can't help feeling that, in some ways, we might be the poorer for it. Some researchers argue, for instance, that our disconnection from geometry stymies our creativity.[19] Our brains now lack a dimension of capability that used to exist in students who actually used geometrical thinking, they say. But the evidence is not overwhelming, and perhaps it doesn't matter. Whatever the truth, geometry's role in shaping our art, architecture, and exploration is beyond dispute. And, betraying my eight-year-old self, I am now convinced that the joy of geometry is just something that everyone should experience, regardless of whether it's useful or changes the brain.

I'm not sure I'd say the same about algebra. Thomas Jefferson once described the study of this subject as 'a delicious luxury',[20] and the English writer Samuel Johnson recommended it as a means to make the mind 'less muddy'. However, others have been less enthusiastic. Even someone as well-versed in algebra as the British mathematician Michael

Atiyah saw it as a double-edged sword. Algebra, he suggested, robs the mathematician of geometry's intuitive connection to the real world; it is mathematics stripped of humanity. 'Algebra is the offer made by the devil to the mathematician,' he once said. 'The devil says: I will give you this powerful machine, it will answer any question you like. All you need to do is give me your soul.'[21]

Is algebra worth this price? We'll soon be in a position to decide.

Chapter 3

ALGEBRA

How we got organized

It's all very well being able to count, but what if some things are unaccounted for? Learning how to create and use mathematical tools like quadratic equations gave us the power to find missing numbers and bring the processes of the natural world under our control. The basics, such as how much tax to pay, or how best to win a battle — even if that battle was only against another mathematician — soon morphed into sophisticated algorithms for solving every kind of problem, from predicting the movements of the planets to cutting the cost of motoring and enabling humanity to survive the Cold War.

On Tuesday, 17 April 1973, a small shipping company ignited a revolution in an industry that didn't know it needed one. Using 14 small aircraft, it delivered 186 packages to 25 cities overnight.[1] The revolution is simple to describe: every single one of those journeys started out in Memphis, Tennessee.

This 'hub-and-spoke' operation began with just 389 employees and took two years to start making money. Today, though, it employs 170,000 people and has an annual revenue of $71 billion. You know it as FedEx.

The success of FedEx stems from its founder's decision to put the hub of its operations at the centre of all possible delivery locations in the US. How do you do something like that? Frederick W. Smith, founder and CEO of FedEx, almost certainly picked up a map of the United States and looked for the airport nearest to the point that would minimise the average distance a package would have to travel. There were other considerations: the airport would need to have good weather all year round so that it rarely closed, and its operators would have to be willing to alter some of the infrastructure to accommodate Smith's business.

As it turns out, Smith could have done better. In 2014, maths professor Kent E. Morrison developed an algorithm to do this more systematically, using census data to find where everyone in America was located.[2] Morrison found that the optimal location was about 70 miles southwest of Indianapolis, at a location in Greene County, Indiana. Smith's hub in Memphis was just 315 miles away. Interestingly, FedEx's competitor UPS had got a little bit closer. UPS had shamelessly adopted the hub-and-spoke system shortly after FedEx made a success of it, and chose Louisville, Kentucky, as their hub, just 275 miles from the central point of the US population.

FedEx and UPS are classic logistics operations. Logistics is all about labelling, sorting, grouping and dispatching. Human civilisation has been built on solving logistics issues, whether it was the ancient Egyptians' construction of the pyramids, Napoleon's realisation that an army marches on its stomach, or the modern-day problem of ensuring airline pilots, Amazon deliveries, and packets of information traversing the worldwide web are exactly where they need to be at any given moment. All of these challenges involve solving a series of puzzles that mathematicians call algebra. So it's fair to say that just getting this book into your hands, whether you have a print version or an electronic file, involved algebra. It's the maths that delivers.

Solving the Quadratic Equation

What even *is* algebra? You might think of it — quite justifiably, given how it has traditionally been taught — as a terrifying labyrinth of equations, an alphabet soup of x, y, z, a, b, and c, plus some superscripts (^{2} and ^{3} and maybe even ^{4}). To the uninitiated, it is certainly off-putting. But there's no reason algebra should be problematic. It's really just the art of teasing out hidden information using what we do know.

Algebra's name comes from the word *al-jabr* in the title of Muhammad al-Khwārizmī's 9th-century book (we met it in Chapter 1 as *The Compendious Book on Calculation by Completion and Balancing*). This pulls together Egyptian, Babylonian, Greek, Chinese, and Indian ideas about finding unknown numbers, given certain others. Al-Khwārizmī gives us prescriptions — formulas we call algorithms — for solving the basic algebraic equations such as $ax^2 + bx = c$, and geometrical methods for solving 14 different types of 'cubic' equations (where x is raised to the power of 3).

At this point in history, by the way, there was no x, nor anything actually raised to any power, nor indeed any equations in what al-Khwārizmī wrote. Algebra was originally 'rhetorical', using a convoluted tangle of words to lay out a problem, and to explain the solution. The sought-after hidden factor was usually referred to as the *cossa*, or 'thing', and so algebra was often known as the 'Cossick Art': the Art of the Thing. An early student of the Cossick Art might find themselves face to face with something like this:

Two men were leading oxen along a road, and one said to the other: 'Give me two oxen, and I'll have as many as you have.' Then the other said: 'Now you give me two oxen, and I'll have double the number you have.' How many oxen were there, and how many did each have?

or

I have a single linen cloth which is 60 feet long and 40 feet wide. I wish to cut it into smaller portions, each being 6 feet in length, 4 feet in width, so that each piece is big enough to make a tunic. How many tunics can be made from the single linen cloth?

These examples were collected by Alcuin of York in around AD 800, and published in a compendium of puzzles called *Problems to Sharpen the Young*.[3] They're not that different from the questions we faced in maths lessons at school.[4] However, we had the advantage of being able to turn them into equations; it is worth pausing, before we go deeper into algebra, to appreciate how privileged this makes us.

It was only in the 16th century that anyone thought to move algebra away from words. The idea came to a French civil servant called François Viète. After training as a lawyer, Viète spent most of his professional life in the service of the French royal court, helping out in any way he was asked to. He was an administrator in Brittany, a royal privy counsellor to Henry III, and codebreaker to Henry IV. Viète's proudest moment might have come when the king of Spain accused the French court of sorcery. How else, he complained to the Pope, could France have foreknowledge of Spain's military plans? But there was no sorcery, of course. Viète was simply cleverer than Spanish codemakers, and had been able to decrypt their communications when French soldiers intercepted them.

It was perhaps this same mental agility that enabled Viète to see that rhetorical algebra would be easier if it was encoded as symbols. In his algebra, he used consonants to designate parameters, and vowels for the unknown items. He would write something like:

A cubus + *B* quad. in *A*, æquetur *B* quad. in *Z*

where we would now write

$$A^3 + B^2A = B^2Z$$

It still wasn't plain sailing, if we're honest, but it was a start. It is interesting to note that the sign for plus is here (and he used minus signs elsewhere), but the equals sign is not. Welsh mathematician Robert Recorde introduced our equals sign in 1557, in his snappily titled book *The whetstone of witte, whiche is the seconde parte of Arithmetike: containyng the xtraction of Rootes: The Cossike practise, with the rule of Equation: and the woorkes of Surde Nombers.*

And while we're on the subject of notation, it's worth noting that the reason that the letter 'x' became associated with the unknown thing is still hotly disputed. According to cultural historian Terry Moore, it's because al-Khwārizmī's original algebra used *al-shay-un* to mean 'the undetermined thing'.[5] When medieval Spanish translators were looking for a Latin equivalent, they used the closest thing they have to 'sh', which doesn't actually exist in Spanish. And so we ended up with the letter that makes the Spanish 'ch' sound: x. But other sources say that it is down to René Descartes, who simply put the two extremes of the alphabet to work in his 1637 book *La Géométrie*.[6] He generalised the known parameters to a, b, and c; the unknowns were designated x, y, and z.

If you are intimidated by the idea of algebra, with all its enigmatic notation, you might benefit from thinking of it as just a way of translating geometric shapes into written form.

In structuring this book, I have drawn an artificial distinction between algebra and geometry. Although we typically learn them as distinct topics — mostly because it makes it easier to design school curricula — algebra flows seamlessly from geometry; it is geometry done without pictures, a move that liberates it and allows the mathematics to flourish. To see how, let's return — as ever, it seems — to the ancient practices of taxation.

As we saw in our look at geometry, taxes were often based on field areas — the Babylonian word for area, *eqlum*, originally meant 'field'.[7] It's no wonder that Babylonian administrators had to learn how to solve puzzles like this one offered up on the ancient Babylonian tablet YBC 6967, which sits in the Yale collection:

> The area of a rectangle is 60 and its length exceeds its breadth by 7. What is the breadth?

Let's try solving it. If the breadth is x, the length is $x + 7$. The area of a rectangle is simply the breadth multiplied by the length, so the area A is given by this equation:

$$A = x(x + 7)$$

The parentheses here tell you to multiply each of the things inside the parentheses by the thing immediately outside it, which leads to:

$$A = x^2 + 7x$$

The Babylonians would solve this via a series of steps that illustrate the close connection between algebra and geometry. The process is known as 'completing the square'.

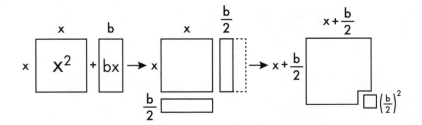

The Babylonian method for 'completing the square' to solve quadratic equations

To make an equation of the type $x^2 + bx$ manageable, you first draw it as geometrical shapes. x^2 is just a square of side x. bx is a rectangle of length x and breadth b. Split that rectangle in two lengthways and move one half to the bottom of the original square, and you can *almost* make a larger square. To complete that larger square, you just need to add in a tiny square of side $b/2$. The area of this tiny square is $(b/2)^2$. So you can see that the original expression is actually equivalent to $(x + b/2)^2 - (b/2)^2$.

Given an equation of the form

$$x^2 + bx = c$$

the Babylonians would substitute in the result of completing the square, making it:

$$\left(x+\frac{b}{2}\right)^2 - \left(\frac{b}{2}\right)^2 = c$$

Then they would work this through and reduce it all down to the formula (though it wasn't written as a formula in the modern sense):

$$x = \sqrt{\left(\frac{b}{2}\right)^2 + c} - \left(\frac{b}{2}\right)$$

The answer is that the breadth is 5, and the length is 12. But I wonder if that formula looks slightly familiar to you? If I offer you a tweak to the original equation so that you have

$$ax^2 + bx + c = 0$$

you would solve this using a formula you learned at school — the quadratic formula:

$$x = \frac{-b \pm \sqrt{b^2 - 4ac}}{2a}$$

As you can clearly see, what you learned at school is little more than a 5,000-year-old tax calculating tool. None of us grow up to be

Babylonian tax officials, though — so why are students learning the quadratic formula these days? It's a fair question, and one that causes arguments even among maths teachers.

The Curves of the Cosmos

In 2003, veteran British maths teacher Terry Bladen suggested, during a trade union conference, that quadratic equations were best left to students who actually enjoy maths.[8] Basic numeracy, he reckoned, was enough for most young people to thrive. So outraged were other maths teachers that one even responded on the political stage. Tony McWalter had spent decades teaching the subject before being elected to Parliament. 'A quadratic equation,' he pronounced to the House of Commons, 'is not like a bleak room, devoid of furniture, in which one is asked to squat. It is a door to a room full of the unparalleled riches of human intellectual achievement. If you do not go through that door — or if it is said that it is an uninteresting thing to do — much that passes for human wisdom will be forever denied you.'[9]

Is this true? Finding quadratic equations difficult hasn't stopped people appreciating human knowledge and wisdom; after all, very few of us have had to use the quadratic formula in anything other than formal examinations. But for those who spent their adult lives far from mathematical endeavours we can still honestly say this: learning to handle algebra increased your ability to think in abstract terms, and to pay attention to something your brain would rather not think about. Millennia of experience and some interesting modern research has taught us that (as with geometry) dealing with abstract varying quantities and the numerical relationships between them actually makes our minds work better.[10] Algebra makes you creative, productive, and tenacious, a lateral thinker capable of pushing through to the end of a line of thought. A good example is the German physicist Georg Christoph Lichtenberg.

In 1786, Lichtenberg wrote his friend Johann Beckmann a rather unassuming letter.[11] 'I once gave an exercise to a young Englishman, whom I taught in algebra,' Lichtenberg said. The exercise in question was 'to find a sheet of paper for which all formats forma patens, folio, 4to, 8, 16, are similar to each other.'

What Lichtenberg is asking is a bit like studying similar rectangles, rather than similar triangles. He wants to know how to find the height to width ratio for a sheet of paper that allows you to halve the largest 'folio' size to get the 'quarto' size, then halve that to get the octavo size, and so on. Lichtenberg was interested enough by the answer his student found that he checked it against the paper stored in his writing desk. 'Having found that ratio, I wanted to apply it to an available sheet of ordinary writing paper with scissors,' he tells Beckman, 'but found with pleasure, that it already had it. It is the paper on which I write this letter.'

And now he gets to the point. Lichtenberg wants to know if Beckmann is in touch with any paper manufacturers — he wants to know how come they are using this format, which, he says 'appears not to have emerged by accident'. Did someone in the paper industry already do the algebra?

We don't know. But this letter, about nothing more than an exercise in algebra and the surprising discovery that a mathematical solution may have evolved naturally, is the root of the European standard for paper sizing. In 1911, the Nobel Prize–winning chemist Wilhelm Ostwald published a call for Lichtenberg's ratio to be used as a global standard in paper production.[12] In 1921 it became the German standard, and quickly spread through Europe. In 1975 it was adopted as the official United Nations document format. You probably know it as the 'A' series. Unless you are in North America, which has never felt the need to adopt Lichtenberg's ratio, you have probably already held a sheet of A4 paper in your hand today. It is an invaluable resource for anyone who needs to maintain proportion — whether you're scaling up an art print, or scaling down a paper aeroplane design.

Lichtenberg's paper size question is effectively laid out as rhetorical algebra. The solution is something we have already met — it's a length-to-width ratio of $\sqrt{2}$ to 1. A single sheet of A0 has an area of 1m^2, which means its side lengths are 1.189m and 0.841m. Hold it in portrait orientation and cut it in half across the middle and you have two sheets of A1, whose length is now the width of the A0 sheet, and whose width is half the length of A0. Do the same with each of those, and you have four sheets of A2. All your sheets have had the same length-to-width ratio. Cut the A2 in half across its width and you have ... well, you can guess the rest.

The algebraic formula that gave rise to this standard is not too hard to deal with. If Lichtenberg's student used 'symbolic' algebra, and called the length of his mythical paper x, and its width y, he needed the ratio of x to y to be the same as the ratio of y to half of x. He could write the equation:

$$\frac{x}{y} = \frac{y}{x/2}$$

and then rearrange it to show that

$$\frac{x^2}{y^2} = 2$$

which means

$$\frac{x}{y} = \sqrt{2}$$

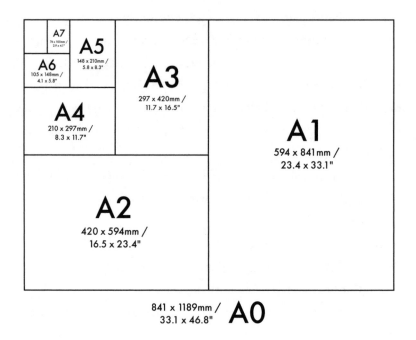

The 'A series' of paper sizes, which all share the same length-to-width ratio

McWalter was right in asserting that algebra has shaped our intellectual achievements; paper size standards provide one of myriad applications of quadratic equations in the real world. They also provide ways of calculating profits for businesses launching new products, and of capturing satellite signals in a parabolic dish. But many of the most useful applications of quadratic equations can be found in describing natural processes, such as the trajectories of the heavenly bodies. To see why, let's look at the curves that quadratic equations create.

If we draw out any quadratic equation on a graph, by calculating the y value for each x value, we will see a line that turns back on itself in some way, creating a curve. Collectively, these curves fall into four classes: parabolas, circles, ellipses, and hyperbolas. If you know what you're looking for, a glance at the equation will tell you what kind of curve it will give. If just one of x or y is squared, it will be a parabola. If both x and y are squared, and the numbers (better known as coefficients) in front of them are the same, you'll get a circle. An ellipse comes from a circle-like equation, where the coefficients of x and y are both positive, but different from each other. A hyperbola arises from an equation where the coefficients of x and y have different signs: one is positive and one is negative.

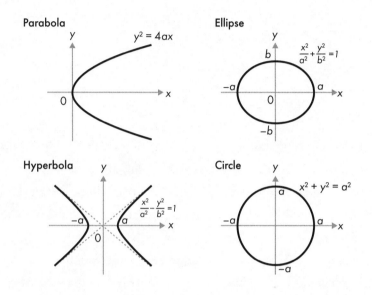

The series of curves defined by various quadratic equations

The various coefficients (a and b in the diagram) determine how elongated the various shapes are — or how large the circle is. Together, these shapes were originally known as conic sections. That's because they arise when something conical intersects a flat plane. If you can, find yourself a torch and switch it on in a dark room to produce what is, effectively, a cone of light. If you shine that cone directly downwards, the cone of light's intersection with the floor creates a circle. This is the simplest of the conic sections. Now point the torch at the wall, at an angle of about 45°. The cone of light intersects with the wall to make an elongated circle: an ellipse. We can also make a parabola by shining the torch at an angle to the wall that is equal to the angle of the cone's sides. Adjust again and you get half a hyperbola.

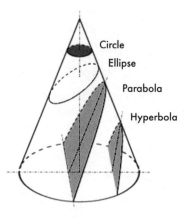

How to cut a cone of torchlight to give quadratic curves

The interesting thing about this is that all these mathematical shapes occur in nature. That's not really news to you, of course — you've seen parabola-shaped rainbows and pictures of Earth's elliptical orbit through the solar system. Nonetheless, it is significant: it means that we should be able to write down equations that describe natural phenomena with mathematics. And that opens a road to deep understanding.

The ancients kept careful numerical records detailing how often the various celestial events happened — comets, eclipses, conjunctions, and so on — and looked for patterns that allowed them to calculate when the next significant moments would be. But right up to (and including) Johannes Kepler in the first years of the 17th century, this was nothing more than number-crunching. Kepler used Tycho Brahe's data to work out that the orbits of the planets are elliptical, but he had no idea why that should be. The Greeks had always said the heavens would be described by the circle, which they considered a perfect shape — who could hope to explain why the universe would be crammed with bodies moving in elliptical orbits? The answer to that question is clear now: anyone who could tie together the planets' observed motion and the idea of a single force acting upon them. Isaac Newton, for example.

We now know, thanks to Newton's pioneering mathematical work, that an object moving in one direction, but also experiencing a single force in a different direction, will follow a path that looks like one of the conic sections. Depending on the object's speed and the force's strength, the movement will be circular, like one of our satellites; or elliptical like the planets' orbits around the sun; or parabolic or hyperbolic, like some of the comets that rush past Earth every now and then. The equation of motion — Newton's equation describing gravitational force, say — is also the equation of the trajectory when it is plotted out over time.

Making sense of planetary orbits was not the first big application of algebra in the West. Unsurprisingly, military minds had already asked whether algebra might solve problems such as finding the angle at which a soldier should set their cannon, given the enemy's position. The answer is yes: algebra can blow that question wide open.

The Art of War

It's not uncommon to hear of researchers wringing their hands over the military applications of their work. Perhaps the most famous example is atomic scientist Robert Oppenheimer, who led the Manhattan Project to create the world's first nuclear weapons. Three years after the first detonation of an atomic bomb, he declared that 'the physicists have known sin; and this is a knowledge which they cannot lose'.[13] In the 16th century, the mathematician Niccolò Tartaglia expressed the same sense of shame.

Tartaglia is actually a nickname meaning 'the Stammerer'. It dates back to Niccolò's childhood in Brescia, when the town was under assault by French soldiers. He was hiding in a chapel with his mother when the French broke in and a soldier's sword slashed across Niccolò's mouth. The wound almost robbed poor Niccolò of his ability to speak, but the Stammerer was a boy of immensely strong character. Not only did he survive (partly because his devoted mother licked his wounds to keep them clean), he also overcame his family's abject poverty and educated himself to the point where he became a respected mathematician.

As part of his research, Tartaglia worked out the 'gunner's question': the relationship between a cannon barrel's angle of elevation and the distance its shot will travel.[14] It gave him a crisis of conscience. This application of algebra was, he said, 'harmful', 'destructive to the human species', and 'a reproachful, vituperative and cruel thing, worthy of heavy punishment by God and by human beings'. And so he burned all his manuscripts.

Then he changed his mind.

Once the Ottoman Emperor Suleyman was on the rise, threatening all of Christendom, Tartaglia's squeamishness about using his new science of artillery to slaughter humans was quelled by a commitment to his Christian brothers and sisters. 'Seeing that the wolf is to ravage our flock,' he wrote to his patron, the Duke of Urbino, 'it no longer appears permissible to me at present to keep these things hidden.' And so he shared the algebra of artillery with the Duke.

What determines the trajectory of an artillery shell? Let's imagine our cannon is at point p on the x-axis, and firing horizontally at a target sitting at q. The shell leaves the barrel at velocity v. We'll ignore air resistance and say the only acceleration acting on it is gravity, a, which means it will trace out one of the conic sections: a parabola. After time t it has reached q, which is at

$$q = p + vt + \tfrac{1}{2}\,at^2$$

In other words, the final position along the x-axis is the sum of the initial position; the velocity times time or t (velocity times time is just distance) and half the acceleration times t squared.

You probably don't want the cannon to fire horizontally, though. And when you elevate the cannon's barrel to an angle A to achieve a specific range, you have to think about the horizontal and vertical components of the shell's flight trajectory. Now we're back in the realm of trigonometry.

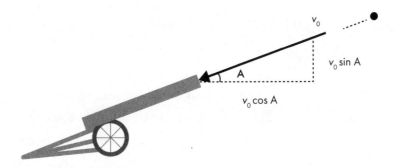

Finding a cannon's range is a matter of triangles

The vertical velocity is $v \sin A$ (where v is the initial velocity). That vertical velocity will reduce to zero when the shell reaches its maximum height. The same force that slowed it down — gravity — will then accelerate it towards the ground, and so the shell begins to fall, and because there are no new forces in play, the fall will take the same time as the rise. The horizontal velocity, meanwhile, $v \cos A$, stays constant (we are ignoring air resistance, remember). Algebra allows us to work out the time that elapses before the shell hits the ground, and what horizontal distance it will cover. If we know how far we want to fire the shell, we can adjust A to give us the perfect range for our shot.

Tartaglia's work was just the first of algebra's military applications. Another is the problem of working out how big your camp needs to be, given the number of soldiers that must be accommodated. Or maybe you want to find out the pay and supplies required for a battalion. There is the simple issue of assigning enough soldiers to dig a trench of a given size in a given time. Or the problem of how much gunpowder a weapon might need. In his 1579 book entitled *An Arithmetical Warlike Treatise called Stratioticos*, Leonard Digges explains how to use algebra to solve all these problems.[15] He warns, for instance, that if you know the amount of gunpowder needed for a weapon half the size of the one you have, you can't just double up. It all has to be worked out using 'the numbers resulting by Cubicall Multiplication', since 'the Rule of Proportion playnlie fayleth'. In other words, if the gun is twice as big, you'll need 2^3

(that is, 8) times as much powder, not simply twice as much. Digges also poses the following question about distributing weaponry:

> There is delivered to the Serjeant Maior 60 Ensignes, in everie Ensigne 160 Pikes, and short weapon. The Generals pleasure, is that he shall put them into one mayne Squadrone, and to arme it rounde with seaven ranckes of Pikes, I demaund how many Pikes, how many Halbers, he shall use to mak the greatest Squadrone, and howe manye Ranckes shall be in the Battayle.

It was a pressing issue of the time: commanders had to know how best to divide and distribute weapons between their various infantry cohorts in order to maximise their effectiveness, while protecting them from potential evisceration during the enemy's cavalry charges. At this point in history, most battles were fought using troops arranged in geometric formations. Getting those formations right was a matter of life and death for the troops, and often critical for a nation's fortunes. The solution to Digges' problem can be found through algebraic searches for the unknown thing. The answer to the first question? 2,520 pikes.

About the same time that Tartaglia was working on the algebra of cannon fire, mathematics was being weaponised in another, very different way. At this point in history, skill with algebra was still rare and impressive enough that it had become a means by which one mathematician could prove their superiority to another, and maybe take their job. The serious consequences of such mathematical duelling — a mathematician could starve if defeated — quickened the Darwinian evolution of new algebraic solutions. It really was survival of the mathematically fittest, but the survivors had to be careful, and professional mathematicians were judicious about which young people they would train in the algebraic arts. The last thing any of them wanted was a student who would share their secrets with competitors, or challenge the teacher in order to take his job. The result was a slow spread of mathematics, and an ever-growing mistrust between mathematicians. We don't often associate mathematics with secrecy, jealousy, and paranoia, but the story of how we went beyond the quadratic equation, solving cubic (x^3) and quartic (x^4) equations, has them all.

The Battle of the Cubic

Our tale begins with a familiar name: Luca Pacioli. In his 1494 *Summa de Arithmetica*, Pacioli declared that, while there was a general formula for solving quadratic equations (the quadratic formula that we looked at earlier in this chapter), it seemed impossible to find a general solution to the cubic equation, where x is raised to the power of 3. In other words, something of the form:

$$ax^3 + bx^2 + cx + d = 0$$

Pacioli's declaration was of purely intellectual interest; there was no application in sight for cubic equations. Nonetheless, Pacioli's one-time collaborator, a mathematician from Bologna called Scipione del Ferro, rose to the challenge. He found a way to solve a related cubic equation, one where b is zero, so that it has the 'depressed' form without x^2:

$$ax^3 + cx + d = 0$$

Like any sensible mathematician of his time, del Ferro divulged his solution to no one. Until, that is, he found himself on his death-bed. Once he knew he wasn't long for this world, del Ferro called his student Antonio Fior and his son-in-law Annibale della Nave to his bedside, and bequeathed them the solution.

These men were of very different character. The son-in-law respected the honour, and told no one he now had access to precious mathematical knowledge. The student, Fior, though, was greedy and ambitious. He saw the solution to the depressed cubic equation as a deadly weapon. His first victim, he decided, would be Niccolò Tartaglia.

In 1535, when Fior issued the challenge, the Stammerer was working in Venice as a teacher of Euclid's theorems. Fior coveted Tartaglia's position, and followed the standard protocol to challenge the Stammerer to a mathematical duel. He and Tartaglia set each other 30 problems. All of the problems Fior had set were variants on mathematical solutions for the depressed cubic equation. Tartaglia saw immediately that Fior must have a solution, and that keeping his job depended on his ability to work out that solution for himself. And, gifted mathematician that he was, he did it. On February 12, Tartaglia found a way to solve the depressed cubic equation $x^3 + px = q$. The following day, he worked out how to solve an equation of the type $x^3 = px + q$. Shortly afterwards, he discovered the path to solving $x^3 + q = px$. Tartaglia was able to solve all Fior's reduced cubic problems. Fior, on the other hand, struggled with the problems that Tartaglia had set. The contest was a triumph for Tartaglia, who kept his job, and burnished his reputation even further by publicly waiving his winner's right to 30 lavish banquets. Like a defeated Rumpelstiltskin, the humiliated Fior disappeared from public life.

There was to be no happy ever after for Tartaglia, however. At the time of the duel, a celebrated Milanese mathematician by the name of Jerome Cardano was in the middle of a grand project: a book detailing all of the algebraic knowledge of his times.[16] Cardano heard about Tartaglia's conquest of the depressed cubic equation and asked him to contribute his solution to the book. Aware of its value, Tartaglia refused. Cardano repeated the request, promising to publish Tartaglia's solution with full credit. Again, Tartaglia refused. Then Cardano offered to put Tartaglia in touch with generals who would pay handsomely for access to Tartaglia's expertise in the mathematics of artillery. Still Tartaglia would not budge. Eventually, Cardano was reduced to a strange offering: if Tartaglia would just let him in on the solution — mathematician to mathematician — he would appreciate it, but not publish it. And, for no discernible reason, this was when Tartaglia finally caved.

Armed with Tartaglia's solution, Cardano and his student Lodovico Ferrari set about developing it into a solution to the full cubic equation. They succeeded — and went even further. Working from Tartaglia's innovation, Ferrari came up with a solution to the quartic equation, where an x^4 term is added in. As with the cubic equation, there was no practical use for the quartic, but Cardano put all of it into his manuscript. However, Cardano couldn't publish it because everything depended on Tartaglia's original solution — which he had sworn not to publish.

The resolution to this impasse came in the form of a schoolteacher from Brescia called Zuanne da Coi. He was an acquaintance of Tartaglia's, and had heard that Scipione del Ferro had shared his solution to the reduced cubic with his son-in-law as well as Fior. Perhaps, da Coi suggested to Cardano and Ferrari, they should all pay the son-in-law a visit? They did, and came away with the secret that Fior had inherited and Tartaglia had worked out. In a move that stirs debate among scholars even today, Cardano published his book, assuring himself that, technically, this did not constitute breaking his oath to Tartaglia.

Tartaglia was outraged that his hard-won (and extremely valuable) solution to the reduced cubic was now available to anyone who bought Cardano's book. The two men exchanged a series of open letters, Tartaglia's prose becoming ever more vitriolic. The aggrieved man demanded satisfaction in a mathematical duel. Cardano, who had the most to lose, refused the challenge. And then a plum job opened up in Tartaglia's home town of Brescia. The Stammerer applied, and was accepted on one condition: that he take on Cardano's student, Lodovico Ferrari, in a public duel.

Ferrari was eager to compete against a man who had repeatedly slandered his beloved teacher. And so the two men exchanged their problems and met in front of a fascinated crowd in the Garden of the Frati Zoccolanti in Milan on 10 August 1548. Unfortunately for

Tartaglia, Ferrari had a deeper understanding of the cubic and quartic solutions than the Stammerer, and had used them to devastating effect in his problem-setting. He gave questions such as:

> There is a cube such that its sides and its surfaces added together are equal to the proportional quantity between the said cube and one of its faces. What is the size of the cube?

and

> Find me two numbers such that when they are added together, they make as much as the cube of the lesser added to the product of its triple with the square of the greater; and the cube of the greater added to its triple times the square of the lesser make 64 more than the sum of these two numbers.

and

> There is a right-angled triangle, such that when the perpendicular is drawn, one of the sides with the opposite part of the base makes 30, and the other side with the other part makes 28. What is the length of one of the sides?

Tartaglia could not solve them all. He left Milan in disgrace. He still got the job in Brescia, but only for 18 months, after which his disappointed employers stopped paying him. Ferrari, on the other hand, became a local celebrity, and was offered a plum job of his own: as the Holy Roman Emperor's chief tax assessor for Milan. Though this kind of algebra was still widely considered of no practical use, Ferrari's algebraic skills — which he may well have never used again — enabled him to retire a very wealthy man.

Let's pause for a moment and think about how we would feel when presented with the depressed cubic equation $x^3 + 6x = 20$. Could you solve it? Cardano's solution presented in *The Great Art* is this:

> Cube one-third the coefficient of x; add to it the square of one-half the constant of the equation; and take the square root of the whole. You will repeat this, and to one of the two you add one-half the number you have already squared and from the other you subtract one-half the same ... Then, subtracting the cube root of the first from the cube root of the second, the remainder which is left is the value of x.

It seems rather daunting, doesn't it? But really, it's just geometry. Cardano's solution begins by imagining a huge cube that he divides into six blocks and smaller cubes — it's essentially a 3D version of completing the square. He knows the dimensions of each one of these shapes, and that their volume adds up to the volume of the big cube of which they are component parts. He reduces this to a quadratic equation and finds a solution: $x = 2$.[17] As you can see, it's easy to check this. That's why the mathematical duels were such fun for the crowds: it was immediately apparent whether or not one of the combatants had succeeded.

In *The Great Art*, Cardano wanted to present a general solution, one that worked for any kind of cubic equation. But that was far from straightforward because he had to work through multiple forms of the equation. He had to deal with things of the form $x^3 + mx = n$ and $x^3 + n = mx$ separately, for instance. Even with quite basic mathematical training, today we would rearrange them so that they were equivalent forms, but with negative values for m or n. But at this point in mathematical history, not only was there no kind of notation that would help in the rearrangement of an equation, there was also discomfort with the very idea of negative numbers. That's why Cardano worked through these two equations in entirely separate chapters (and why Tartaglia developed separate solutions for $x^3 = px + q$ and $x^3 + q = px$).

Eventually, though, the completing-the-square-style trick of substituting an expression for x gave a general solution for an expression that we would now write as

$$ax^3 + bx^2 + cx + d = 0$$

The trick turns it into an equation that can be solved using Cardano's formula for the depressed cubic. *Ars Magna* details how Ferrari found a similar innovation for solving quartic equations — those involving x^4 terms.

And what of the quintic equation, where a forbidding x^5 term comes into play? Cardano and Ferrari suggested it might be possible to solve it using the same old trick of making a substitution for x. But, they eventually admitted, they hadn't been able to find such a method. They were right to give up the hunt. Almost three hundred years later, in 1824, a Danish mathematician called Niels Abel showed that a solution via substitution was impossible.

There is a way to solve the quintic equation, as it turns out. It uses something called elliptic functions (or elliptic curves) that are now used in cryptography, the science of hiding secrets. We'll get to that in a later chapter; for now, let's look at the modern-day applications of these quadratic, cubic, and quartic equations. At first glance, it's a continuation of Tartaglia's work. But where the Stammerer described the curves traced out by the paths of cannonballs, more recent innovators have focused on making curves in physical objects — the Ford Taurus, for example. And it's here that we start to see how algebra still helps to solve some of our most pressing problems, even in an advanced, technological society.

Curving the World

In 1974, a gallon of petrol cost Americans around 40 cents. In 1981, putting the same amount of fuel in the tank cost $1.31. US car manufacturers realised they had to do something to save motoring. But what? Redesigning their engines to be significantly more efficient would be too difficult. Much better to make the cars more aerodynamic.

The first truly aerodynamic American family car was the 1986 Ford Taurus. It's hard to appreciate now, but this looked radical to the American eye — so radical, in fact, that Paul Verhoeven chose it as Robocop's car in his 1987 tale of a future police force experimenting with a cyborg officer. The Taurus was already a year old, but it still looked like a car of the future. Why? Because it had curves. Until the Taurus, American cars of the time were best described as boxy. Their bodies' straight lines were easy to manufacture, and the fact that the boxiness increased drag and reduced fuel efficiency hadn't been a problem: petrol was cheap. Across the Atlantic, however, things were different.

1986 Ford Taurus
IFCAR, Public domain, via Wikimedia Commons

In Europe, taxation had always elevated the price of fuel. In the 1970s, rising oil prices meant that Europe's motorists had long been facing high running costs for their vehicles. They had found a partial solution to the problem, however: curvy, aerodynamic body shapes.

Luckily for Ford, the company had just repatriated a Europe-based American designer called Jack Telnack. During his years on the other side of the ocean, Telnack witnessed the evolution of the fuel-saving curves we now associate with European car design first hand.

Ford's engineers couldn't just plug the equation for the curve they wanted into a steel bending machine. Neither could they give a computer thousands of points to plot for every curve in their design — it's just too inefficient. Their engineers had had to find another way to produce the required curve. But, as Telnack knew, by the early 1960s, two French car designers had already solved this problem.

Although they are widely known as Bézier curves, Renault's Pierre Bézier and Citroën's Paul de Casteljau should probably share the credit for this innovation. In fact, de Casteljau did most of the mathematical work. But it was Bézier who put it to work in the machine room and made it possible for others to follow their lead.

Bézier curves are best explained by starting with a pair of lines that are two sides of a triangle. Draw them wherever you like on a piece of paper, at whatever angle you like to each other. Make one *AB* and the other *BC*. Now divide them into the same number of sections — say, ten. Number them from *A* = 0 so that 10 comes at *B*, then start again with *B* = 0 and *C* = 10. Now draw a series of straight lines joining 1 to 1, 2 to 2, and so on.

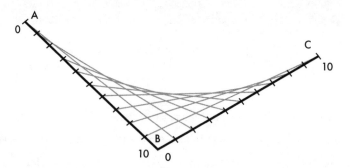

A Bézier curve constructed from straight lines

Can you see the curve? It's not really there; you have only drawn straight lines. But each of your straight lines is a 'tangent' to the curve — that is, a line that touches at only one point on the curve, whose exact shape is defined by the relative positions of A, B, and C.

Bézier called point B the 'control point' because as you move B, you get a different curve. With just one control point, the curve is always definable by a quadratic equation that contains the values of A, B, and C. If you add in another control point, the resulting curve represents a cubic equation. Add another, and you'll have a quartic curve. If you don't want to add control points, you can add curves instead. In much the same way that Cardano and Ferrari found the solution of a quartic equation by reducing it to a cubic equation (and the cubic to the quadratic), you can draw a cubic Bézier curve via the interactions of two quadratic curves, and a quartic curve by the interaction of two cubics.

The curvy Ford Taurus was launched to critical acclaim and — more importantly, given Ford's catastrophic downward trajectory in the automobile marketplace — huge sales. It's not too much of a stretch to suggest that algebra saved the American car industry.[18]

This is not only how the modern aerodynamic car got its shape. This same shorthand for creating any curve you want is useful for building bridges, buildings, and aeroplanes. But it's also handy in less obvious places — such as type. This book, whether you're reading it on paper or on an e-reader, exists through algebra. When you use a Truetype font such as Times New Roman, Helvetica or Courier, you are creating quadratic Bézier curves that define where the ink or pixels are placed.[19]

Perhaps it shouldn't come as a surprise to learn that, if designers are using algebra to give shape to objects in the real world, they also use it to shape virtual worlds. Some are the same computations: video game designers, for example, have to program in quadratic, cubic, and quartic equations in order for their virtual worlds — and the weapons fired within them — to have realistic properties. Architects also have to follow algebraic rules when creating designs that minimise wasted space and optimise room proportions. And entrepreneurs use quadratic functions to optimise pricing and inventory when launching a new product. None of this goes much deeper than what Cardano, Tartaglia and Ferrari could manage — and much of it is now automated within computer software — but it is still there, giving shape to our environment and our experience of the world.

Now let's go a little deeper, because algebra can also help organise things that lie beyond the observable properties and behaviour of a world's contents. It turns out that the hidden structures of our universe can be described through algebra — which is why physicists are so smitten, like the ancient Greeks, with the idea that our universe is mathematical at heart. This field of mathematics is known, somewhat oddly, as 'abstract algebra' — as if the algebra we've been looking at isn't already abstract enough. To be fair, it's sometimes called *modern* algebra. But even that 'modern' label seems a little suspect too. After all, it began with a young Frenchman called Evariste Galois, who died in 1832.

Galois, Noether, and the Algebra of the Universe

'Don't cry, Alfred! I need all my courage to die at twenty!' These have been reported as Galois's last words, spoken to his younger brother. He had sustained a fatal injury in a pistol duel. His opponent seems to have been a competitor for the affection of a young woman called Stephanie — historians' best guess is that she was the daughter of Galois's landlord.

Galois was buried in the commoners' ditch in the cemetery at Montparnasse in Paris, his life barely begun. Nonetheless, he has achieved everlasting fame. Galois is now celebrated as the father of group theory, a branch of mathematics that can be thought of as the zoological classification of algebraic routines. Just as a biologist will lump certain organisms together as mammals, or fungi, or bacteria, mathematicians classify algebraic expressions into groups that share common properties, such as equations that, like the quadratic equation, can be all be solved by the same general method.

Establishing the details of biological classification made it easier to see the bigger picture: it's what gave us the theory of evolution by natural selection. Algebraic classification is no different. It has enabled us to learn huge truths about the universe, such as the structure of the 'particle zoo', the fundamental entities that come together to create all physical matter. Having started with Galois, this endeavour reached its conclusion in 2012 with the discovery of the Higgs boson at CERN in Geneva.

To understand something of how algebra achieves such profound impacts — and we will only aim to understand it in its sketchiest form — let's imagine we've solved a cubic equation to find its three solutions: a, b, and c. Now let's try to find how they relate to each other. We can construct expressions such as

$$(a - b)\,(b - c)\,(c - a)$$

and maybe swap the solutions around. Let's replace a with b, b with c, and c with a. Now you have

$$(b - c)\,(c - a)\,(a - b)$$

This is effectively the same as swapping the order of the bracketed expressions. Calculate the value, and you'll get the same result.

What happens if we only swap a and b? Then we get

$$(b - a)\,(a - c)\,(c - b)$$

This has the same effect as multiplying the original expression by -1: positives have become negative, and vice versa. Which means that if I did this transformation, then squared the result, I would get the same result as if I squared the result of the original expression (as Brahmagupta said when introducing negative numbers, a minus times a minus is a plus).

Galois made a series of general observations of this sort, which allowed him to group together certain sets of algebraic expressions, choosing the groups by examining the ways in which the solutions related to each other under many different kinds of transformations. It might not sound like much, but it has become a central pillar of mathematics.

Galois clearly knew something of the value of his discovery because, according to legend, he spent the night before his duel pulling together all of his papers so they could be passed to his friend Auguste Chevalier.[20] He was apologetic about the rushed nature of the communication. 'Later there will be, I hope, some people who will find it to their advantage to decipher all this mess,' he wrote. His humility only adds to the tragedy of his early death.

Galois's insights were so valuable because the transformations are an abstract mathematical link to the physical trait of symmetry. Swapping a and b in the expression above, for instance, is not unlike the swapping of left and right that occurs in a mirror image.

Symmetry is about changing something, and seeing whether there is a change in its appearance or behaviour. If there is no change, you've found a symmetry. If there is a change, the symmetry can be described as broken. In geometry, we can find simple examples: a square has reflective symmetry along its diagonal, for example. If you were to hold up a flat mirror along the diagonal, you would see the same shape as the full original shape. A square also has four rotational symmetries, each obtained by rotating it through multiples of 90 degrees. If you rotate through just 45 degrees, it looks different (more like a diamond): the symmetry is broken.

In particle physics, where symmetries are described using abstract algebra, the changes involved are a little more complicated. You might swap a particle for its antiparticle, for example. If you see no difference in how their interactions happen, that's a symmetry. A good example of this is when you change the charge on two electrons to its opposite. Two positrons repel each other in exactly the same way as two electrons do. That's charge symmetry.

Symmetries are at the heart of our understanding of the physical world, because many of the processes of physics can be expressed in terms of reflections, rotations, or simple swaps. Those symmetries might be in space, or in time, as well as in physical properties like electrical charge. There is a deep connection between symmetry and the conservation laws that say certain properties of a physical system cannot simply disappear. Take the law of conservation of energy, for example. You might remember from your physics lessons at school that you can convert energy from one form to another — from kinetic to potential, say, by rolling a boulder to the top of a hill — but that energy won't ever disappear from the universe. Some might be dissipated as sound when the boulder crashes down the other side of the hill, and some will be reconverted into kinetic energy — of the boulder, but also of the soil and rocks of the hill as they are kicked by the rolling boulder. But it won't cease to exist. That's because of a symmetry in the laws of physics: quite simply, they are symmetric in time and don't change from minute to minute or even from millennium to millennium. Other symmetries provide other conservation laws. The orbits of the planets around the sun, for instance, have a rotational symmetry that is linked to the conservation of angular momentum. Galois didn't give us this set of insights, though. This was the work of a remarkable mathematician called Emmy Noether.

It appals me that Amalie 'Emmy' Noether is the first woman we are encountering face to face. The truth is, the prejudice encountered by women in mathematics is so deeply ingrained that Noether almost didn't make it in mathematics at all. Her father was a mathematics professor and both her parents were keen that all their children follow an academic path. But it was much easier for Noether's brothers.

Emmy Noether was born in Erlangen, Germany, in 1882. She was fantastically bright, but when she was ready to go to university, she found her path was blocked. Her father's university, Erlangen, didn't yet accept women. Eventually, she was able to gain both undergraduate and postgraduate degrees, but she then found herself stuck once again: no university would offer her paid employment to research or teach mathematics.

Such was her love for the subject that Noether taught without pay for seven years at Erlangen. It was only when her brilliance came to the attention of Germany's premier mathematicians that she found a way forward: David Hilbert and Felix Klein invited her to come and work at their mathematical institute in the University of Göttingen. But, despite their eminence, they couldn't persuade their university to pay Noether either. For four years Noether worked, without a salary, as Hilbert's assistant. It was only in 1922 that she finally secured a paid position at Göttingen. By that time she had already done some of her best work — in fact it was some of the best work ever done in mathematics.[21]

If you want to get a sense of just how good a mathematician Noether was, consider this. After her untimely death at the age of 53, due to complications arising in a surgical procedure, Einstein declared in *The New York Times* that 'Fräulein Noether was the most significant creative mathematical genius thus far produced since the higher education of women began'.[22] It's a demeaning compliment, really: Noether was arguably the greatest algebraist in the world at the time of her death — male or female. What's more, Einstein knew it; when he was stuck on one part of his general theory of relativity — which Noether had helped to fix — he wrote to David Hilbert, asking him to 'charge Miss Noether with explaining this to me'.[23]

Noether's theorem is a way of working Galois's algebraic groups — and much else that had been discovered about algebra since — into an ambitious scheme of classification and categorisation. It was as if everyone else in the field had been feverishly studying their little part of the whole, and Noether strolled in and saw how all the contributions tied together, where they would complement each other, and how to weave a complete fabric from all the threads. It even had deep implications for other fields, such as topology, the mathematics that describes what happens to the properties of geometric shapes when they are twisted and stretched. In a 1996 lecture, the German topologist Friedrich Hirzebruch said that, although Noether barely entered the field, 'she published half a sentence and has an everlasting effect'.[24]

Noether's abstract algebra means that we can use her equations to look for new laws, particles, and forces of physics. When symmetry is broken, there is always a reason, and usually it's because of a force. In fact, this has typically been how physicists discover the existence of an as yet unknown force of nature. In the early 1960s, for instance, the physicist Murray Gell-Mann spent some time examining the symmetries in the abstract algebra that describes the atomic nucleus. He found that the symmetries suggested that there should be more fundamental particles than the protons and neutrons in the nucleus of an atom. In 1964, he published a paper predicting the existence of as yet unknown particles that come together to make protons and neutrons. He named them after a word he liked the sound of when he had read James Joyce's *Ulysses*. Experiments soon found Gell-Mann's 'quarks', and Gell-Mann soon had a Nobel Prize to his name.

It was also Noether's abstract algebra that led Peter Higgs and colleagues to notice, in the 1960s, that an as yet undiscovered particle should be lurking in the depths of particle physics. The elusive 'Higgs boson' was finally discovered in 2012, and Higgs too received a Nobel Prize.

The Higgs boson was the last piece in the puzzle of particle physics; it turns out that the entire set of particles can be conjured from considerations of symmetry and conservation laws as laid out in Noether's abstract algebra. It might have started as a taxation tool but, thanks to algebra, we now know how the universe works.

How to Get Whatever You Want

Let's take a look at a piece of algebra that you may already have used today. This story starts in 1998, when two Stanford University computer science students published a paper that contained the following text in its introduction:[25]

> Automated search engines that rely on keyword matching usually return too many low quality matches. To make matters worse, some advertisers attempt to gain people's attention by taking measures meant to mislead automated search engines. We have built a large-scale search engine which addresses many of the problems of existing systems.

The authors, Sergey Brin and Lawrence Page, called it Google, 'because it is a common spelling of googol, or 10^{100} and fits well with our goal of building very large-scale search engines'. Within a very short time, their very large-scale search engine took over the world. Indeed, by 2006, to 'google' had entered the *Oxford English Dictionary* as a verb describing a common means of finding information.

Google's PageRank algorithm was a masterful application of what is known as linear algebra.[26] This is algebra where the variables (data about internet pages, in this case) tend to be processed in ways that don't depend on the square, the cube, or any other power. So $y = 4x$ would be an operation in linear algebra; $y = 4x^2$ would not.

Linear algebra is an old idea. We have linear algebra texts from China that go back beyond the 2nd century BC. They explore solutions for sets of equations that together contain all you need to find out the relationships between the variables. These 'simultaneous' equations come in what the Chinese texts call 'arrays'. Modern linear algebra uses all kinds of technical expressions that might make you feel uncomfortable: vectors and matrices, for instance, and eigenvectors and eigenvalues. Indeed, the power behind Google's throne has been described as 'the

$25,000,000,000 eigenvector'. But all we really need to know about linear algebra is its equations are, essentially, mathematical spreadsheets where a single operation can process a whole, huge array of data.

The Google algorithm is just one of myriad powerful applications of linear algebra; it's safe to say that the world you know wouldn't work without it. Processing lots of variables simultaneously, and finding out the relationship between them all so you can optimise one particular outcome of their actions, is something we do well now. As well as Google searches, it also powers airlines: flight schedules, fleet plans, aircraft routes, crew pairings, gate assignments, maintenance schedules, food service plans, training schedules, and baggage handling procedures are all optimisation problems solved using linear algebra. Remember to be grateful for algebra when you're eating your next in-flight meal.[27]

FedEx and UPS use linear algebra to search for the optimal delivery routeing programs. Then there's your shopping and its logistics: how does your supermarket get all its produce to the store, and — if you do online shopping — how does that produce get to your home? Through linear algebra. The same is also true in health care: scheduling operations, surgeons, appointments, and deliveries of medicines is a modern equivalent of working out how best to organise the pikes and halberds at your disposal. Even the way your Google search results are delivered to your computer screen — the logistics of routeing information through the internet — is reliant on linear algebra. To be sure, most of this is now written into software, just as trigonometry is programmed into architects' computer-aided design packages. But it's still part of your everyday routines. The unparalleled ease of modern life owes a huge debt to the mathematicians who created algebraic solutions for almost all of our logistical problems.

Linear algebra can even claim to be the sole reason humanity reached the 21st century without destroying itself. The Cold War, 44 years of fragile but largely peaceful standoff between the United States and the Soviet Union, was in large part a product of this area of mathematics.

As the Second World War ended, and relations between America and the Soviets deteriorated into quiet threats of nuclear destruction, groups of mathematicians on both sides dedicated their careers to finding ways to ensure the threats were never realised. The most famous of them was John Forbes Nash, the subject of the biography *A Beautiful Mind*, which was turned into an Oscar-winning film starring Russell Crowe. The film focuses on the deterioration of Nash's mind, and how his mental breakdown affected his family and his career. But sadly lost to the camera's focus is the role of his work — and the work of countless others — in keeping us from the abyss of all-out nuclear war.

You'll probably have heard the phrase 'mutually assured destruction'. It makes avoiding nuclear war sound simple: if both sides just gather enough nuclear weapons — themselves a product of abstract algebra, by the way — no one will want to strike first, because the retaliation, and the rounds of strikes that will follow, will make the planet uninhabitable. But it was far more complicated than that.

The algebra at work here is encapsulated in a field of research known as game theory. It sounds frivolous, perhaps, so let me give you a sense of how seriously the mathematicians behind game theory were taken. In an era when no one was allowed to meet their counterpart behind the Iron Curtain, both sides knew that the chances of mutual destruction would be reduced if these mathematicians were allowed to talk to one another. In 1971, an unprecedented gathering of game theorists from America, Europe, and the Soviet Union took place in Vilnius, Lithuania. This was within a year of the signing of the Treaty on the Non-Proliferation of Nuclear Weapons by the Soviet Union, the United States, and others. All sides wanted to maintain the peace, and allowing their mathematicians to meet was a central part of the strategy.[28]

It would be impossible to describe their mathematical contributions here. Many are difficult enough that even undergraduate maths students aren't taught them in any detail. Some are to do with working out the best response to a threat, given all the mitigating circumstances. Others are to do with the optimal sizes of nuclear stockpiles, given a mutual distrust. Still others focus on how best to invest in, and deploy, countermeasures.

Over the years, hordes of mathematicians have worked on the algebra of the arms race, but John Nash stands above them all. That's because of his celebrated 'Nash equilibrium', an algebraic way to find the best resolution to a dilemma where two parties cannot trust each other. It uses a complex form of linear algebra, and describes a scenario where two opposed parties can find themselves in a situation where they cannot do better than their current standing. The equilibrium strategy might not be the optimal strategy for that player, but it is the only one that doesn't make things worse. In a Nash equilibrium, neither side is happy, but neither side is going to do anything about it: it's a least-worst-option kind of thing.

Finding the conditions for a Nash equilibrium to exist, and the strategies to reach it, won John Nash the Nobel Prize for economics and a rather undervalued place in history. Nash gave both sides in the Cold War concrete proof that they should accept the frosty détente and not consider any further moves. Essentially he used algebra to make the world a safer place. I can't help thinking Niccolò Tartaglia would approve.

Fermat's Last Theorem

Just before we leave this subject, I'd like to reassure you that sometimes even the simplest-looking algebra leaves professional mathematicians floundering. Perhaps you've heard of Fermat's last theorem? It's very simple to describe, but it took hundreds of years for anyone to find a solution.

Pierre de Fermat, who died in 1665, was a French mathematician. Although he was a great thinker, he refused to publish any of his work. After Fermat's death, his son Samuel decided to collect together all his father's papers and publish the notable results. Flicking through his father's copy of Diophantus's *Arithmetica*, Samuel came across a note scribbled in Latin in the margin. Translated into English, it said: 'It is impossible for a cube to be the sum of two cubes, a fourth power to be the sum of two fourth powers, or in general for any number that is a power greater than the second to be the sum of two like powers. I have discovered a truly marvellous demonstration of this proposition that this margin is too narrow to contain.'

Nowadays, we write Fermat's statement like this: given the equation

$$x^n + y^n = z^n$$

there are no solutions for x, y and z when n is larger than 2 — if, that is, the solutions cannot be zero and must be integers.

Fermat made plenty of other scribbled claims of being able to prove a theorem, and the mathematicians who went through his papers were eventually able to find all the proofs in various places — except the one referring to Diophantus's equation. That is why the problem came to be known as Fermat's last theorem.

You can see that, if $n = 2$, we have the Pythagorean triangle of sides 3, 4, and 5 as a set of solutions, because $3^2 + 4^2 = 5^2$. Can it really be so hard to find other solutions? It's a question that Andrew Wiles, the mathematician who finally solved Fermat's last theorem, had been asking since 1963, when — aged 10 — he found a copy of a book about the problem in his local library. 'I knew from that moment that I would never let it go,' he said. 'I had to solve it.'[29]

It took Wiles until 1995, often working obsessively, alone, and in secret. Wiles is now a mathematical celebrity, but there is one question about Fermat's last theorem that even he can't answer: did Fermat really have a proof for it that has been lost?

We know the proof would have been different from the one that Wiles found. The mathematical techniques that Wiles used simply didn't exist in Fermat's time. So if Fermat did have a proof, it uses some mathematical trick that was discoverable in the 17th century, but hasn't been rediscovered since. It seems unlikely, doesn't it? But, that said,

Fermat's other claimed proofs all turned up. Why should this one be imaginary?

In many ways, Fermat's last theorem only scratches the surface of how difficult algebra can be. Mathematicians can create far more equations than they can solve, which is why the entire canon of theoretical physics is not based on actually *solving* the algebraic equations that describe the way that forces and particles interact in the universe, but on finding approximate, good enough solutions. Edward Witten, one of the greats of contemporary physics, once described quantum field theory, our central mathematical description of the universe, as 'a twentieth century scientific theory that uses twenty-first century mathematics'.[30] What did he mean? He meant that it might take us the rest of this century to develop the algebra necessary to understand the universe. Algebra may have been around for thousands of years, but it's far from finished.

Given how useful it has proved, perhaps that's not a bad thing. As we have seen, algebra has already given us a means to solve myriad logistical puzzles, from showing us the best way to arrange a battalion's tents to delivering an algorithm for world peace. It is a spotlight for finding things, be they elusive particles within the theories of physics or documents on a web server. Algebra has settled questions of taxation, got us away on holiday, and exposed the orbits of the heavenly bodies. So who knows, as our exploration of the subject continues, what the algebra of the future will bring?

While we wait to find out, we can at least admire the finished product known to the world as calculus. We invented, developed, and applied this branch of mathematics with astonishing speed, transforming the way we deal with things that move and change. It was essentially completed within a century, and has since brought us revolutions in science, medicine, finance, and — of course — warfare. In fact, it could be argued that calculus was pivotal in bringing America into the Second World War. But, as we'll see, calculus has been centre-stage in numerous battles, right from its very first moments.

Chapter 4

CALCULUS

How we engineered everything

There are still disputes over who invented it, but there's no doubt that calculus changed the world. By harnessing the power of the infinitely big and the infinitely small, calculus brought America into the Second World War and powered the rise of the global financial system. It also enabled the construction of cities, bridges, and climate forecasts.

Its value is simple: it gives us the ability to predict what seems unpredictable. Plenty of people have recognised this and put it to work. The novelist Leo Tolstoy, the designers of the Spitfire fighter plane, the medical researchers who halted the HIV epidemic, and Albert Einstein himself are just a few of those who have flown on the wings of calculus.

In July 1940, a Gallup poll asked American citizens if they would support going to war against Germany and Italy. Eighty-six per cent said no. By September, even after the first peacetime draft in American history, that faction had dropped to 48 per cent.[1] What had Americans learned between July and September? That Hitler was not invincible.[2]

They had learned this from the Battle of Britain, which took place in the skies over England and the English Channel in August and September. In fact, the Royal Air Force fought such an impressive campaign against the German Luftwaffe that American journalist Ralph Ingersoll made the dangerous journey across the Atlantic to London to find out what was happening on the ground. Once back home, he described encountering a 'civilian population holding on despite almost continuous and uninterrupted terror, with only their almost unbelievable courage and faith'. Ingersoll praised the British to high heaven:[3]

> It is no wonder that Adolf Hitler stamps and raves at the psychopathic British. It must be difficult for such a coward to understand such courage. It's extremely difficult for anyone to understand it. But it happened. The Londoners stuck it out without panic, each day burying their dead and binding up their wounded, each day going back to their businesses, unloading their boats in the middle of it all on the bombed docks, opening their stores, putting out their fires and repairing their telephone lines and water mains, shoveling out their streets, going back to their factories.

In the end, he says, they won a victory that would resonate for ever. 'The battle that was fought in the air over London between September 7 and 15 may go down in history as a battle as important as Waterloo or Gettysburg,' he said. What he almost certainly didn't appreciate was that this was a victory born out of calculus.

Calculus is the branch of mathematics that deals with things that change. Whether you are engineering a financial gain, a bridge, a mission to Mars, a war machine, or a glimpse into the future of the planet, calculus is the tool you can't do without. When your maths teacher introduced it, they may well have started with Isaac Newton and the motion of the planets. We, however, will start with Poppy Houston and the motion of an aeroplane — specifically, the Supermarine Spitfire.

In 1931, Supermarine was a British aircraft company high on ideas and low on cash. Their main output was seaplanes, and they had built a sterling reputation on the back of two consecutive wins in a prestigious seaplane race known as the Schneider Trophy. Unfortunately, the British government had just pulled their funding.

This was the era of the Great Depression, and Prime Minister Ramsay McDonald had other priorities; there was no money to invest in the super-fast single-wing aircraft that Supermarine wanted to develop. Enter Lady Houston, England's richest woman.

Born in Lambeth, east London, in 1857, Poppy Houston was a working-class woman — a high-kicking chorus girl at one point — who caught the eye of a succession of wealthy men. She married three of them, and by 1931 two had died, each leaving her with a sizeable fortune. On occasion she was of a mind to put her money to good use. It was she, for instance, who put up the bail to get the suffragette Emmeline Pankhurst out of prison. And when she heard that Supermarine could no longer afford to compete for the Schneider Trophy, she bailed them out too.

Lady Houston was a fiery character. She had ripped up one husband's will when he told her it would provide her with £1 million on his death. This paltry amount, she said, made her feel worthless — a statement that provoked him to raise her inheritance to £5 million. The gift to Supermarine of £100,000 — equivalent to a few million pounds today — was not about high-minded benevolence, Lady Houston said. Nor was it about her garnering prestige as Supermarine's sponsor. It was about securing Britain's military capabilities. Decrying the government's miserly attitude, she declared angrily that 'Every true Briton would rather sell his last shirt than admit that England could not afford to defend herself'.[4]

Lady Houston's cash went towards building the Supermarine S6. This elliptical-winged seaplane evolved, with the substitution of wheels for pontoons, plus a few other changes, into the iconic, Battle of Britain–winning Supermarine Spitfire.

The Spitfire's chief designer, Reginald Mitchell, was built of similar stuff to Lady Houston, it seems. On hearing the name his bosses had decided upon for his new aircraft, Mitchell declared it 'bloody silly'. When prototypes were being tested, he warned the pilot about his team's tendency to offer impenetrable advice. 'If anybody ever tells you anything about an aeroplane which is so bloody complicated you can't understand it, take it from me: it's all balls,' he said.[5] And, perhaps most famously, he expressed utter disdain for the purist's approach to the shape of the Spitfire's wings. 'I don't give a bugger whether it's elliptical or not!' he blustered to Beverley Shenstone, the Canadian-born aeronautical engineer responsible for the wing design. Shenstone did care, though — because he knew about the maths.

In 1907, just four years after the Wright brothers had made the first powered flight, a mathematician called Frederick Lanchester had shown that a wing creates spirals of air called vortices on its rear edge.[6] These, in turn, create a force known as 'induced drag' that pulls backwards on the wing. What's more, the induced drag increases at low speeds and when the aircraft is climbing or diving. That means it affects manoeuvrability. Lanchester noted that, in general, birds have evolved a wing plan that has a suggestion of an elliptical shape, narrowing towards the tip, and that this shape should reduce the induced drag force. He didn't do the maths, but others did. By 1918, aeronautical designers had seen mathematical proof that a 'double elliptical' wing — that is, with elliptical leading and trailing edges — would suffer the least induced drag. Shenstone knew that, if you wanted a fast, agile aircraft, elliptical wings were the way to go.[7]

Shenstone had spent his youth designing boat hulls in Toronto, Canada.[8] His hobby had gradually become a passion, then a career that branched out, via an undergraduate degree in engineering and a master's degree in the design of flying boats, into the burgeoning world of aircraft. By the time he was 23 he had scored a job at Junkers in Germany, where he immersed himself in the theory of flight. In 1931, he returned to England and discovered that, if he really wanted to design world-changing aircraft, he would need to immerse himself in calculus.

The Maths of Change

Calculus is perhaps the most ubiquitously applicable innovation in all of history. You might doubt that, thinking the wheel must be the most widely used invention of all time. However, it's just not the case. The wheel has very limited uses. In fact, calculus has actually been applied to (and improved) every wheel-based technology. What's more, calculus was intimately involved with transport technologies that superseded the wheel, such as the aeroplane and the space rocket. In terms of its impact on civilisation, calculus outguns anything you care to mention — including guns. If you want to calculate the yield of a nuclear warhead, for instance, you'll use calculus.

Whenever there is a continuously changing set of parameters, calculus steps up. Take a jumbo jet's fuel requirements, for example: the amount of fuel needed to keep the plane in the air changes as it burns more fuel and reduces its own weight. Or how to calculate the annual income that can be derived from a savings account that has a variable interest rate. Or the market price of grain as supply and demand fluctuate. All these examples use calculus. And you can even, as Tolstoy did, use calculus as a metaphor. Look at this abridged passage from Tolstoy's *War and Peace*:

> The movement of humanity, arising as it does from innumerable arbitrary human wills, is continuous.
> To understand the laws of this continuous movement is the aim of history. But to arrive at these laws, resulting from the sum of all those human wills, man's mind postulates arbitrary and disconnected units ... Historical science in its endeavor to draw nearer to truth continually takes smaller and smaller units for examination ...

Only by taking infinitesimally small units for observation (the differential of history, that is, the individual tendencies of men) and attaining to the art of integrating them (that is, finding the sum of these infinitesimals) can we hope to arrive at the laws of history.

Tolstoy is using the principles and language of calculus to make sense of history.[9] Integral calculus, for instance, is about adding tiny, *infinitesimally* small bits together — integrating them into a whole, in other words. There is also *differential* calculus, where we infer the laws governing a *continuous* system by calculating the effect of a change in *smaller and smaller units.*

These are all words and practices that calculus pioneer Isaac Newton would have known and recognised. So how did what is widely considered to be the greatest novel of the 19th century end up containing centuries-old mathematical principles? Partly because one of Tolstoy's closest friends was a mathematician. But mostly because the principles of calculus are almost infinitely beguiling to anyone who likes to think about how change happens.

Tolstoy begins his commentary on calculus by introducing the reader to a paradox about motion put forward by Zeno of Elea. Zeno's paradox has come to us in many different forms; Tolstoy examines the one involving Achilles and the tortoise. Achilles and the tortoise are in a race, but the tortoise has a head start. Achilles, though, travels at ten times the tortoise's speed. And still, says Zeno, Achilles can never catch the beast.

The reason is simple. Achilles takes a certain amount of time to cover the distance to the tortoise. And in that time, the tortoise has moved — only by one-tenth of the distance Achilles has covered, granted, but the creature is still beyond Achilles' reach. Now Achilles has to cover that remaining distance, but the tortoise has moved again by the time he does so. 'By adopting smaller and smaller elements of motion we only approach a solution of the problem, but never reach it,' Tolstoy says. In other words, it looks as if Achilles will never actually catch the tortoise.

Of course it's absurd. Achilles would definitely catch — and overtake — the tortoise. The problem with Zeno's argument, Tolstoy explains, is that it doesn't allow for the infinite. 'Only when we have admitted the conception of the infinitely small ... do we reach a solution.' When you can divide the steps of the continuous motion into infinitesimally small chunks (that is, as small as it is possible to be without being zero), and have allowed for an infinite number of steps, Achilles does catch the tortoise. This infinite division is what calculus is all about.

To Infinity

Since the concept of the infinite lies at the heart of calculus, we should pause to give it some attention. The most important thing to note is that infinity is a concept, not a number. As countless playground arguments have shown, there is always a bigger number than any number you care to mention. Infinity is a sort of shorthand for the ultimate point of a sequence that never ends.

That said, infinity is still part of the mathematical numberscape. There is an infinite number of natural numbers (0, 1, 2, 3, 4 ...), for example. There is also an infinite number of even numbers. And there is an infinite number of odd numbers. I'm sure you won't feel there's anything intrinsically odd about that. What is strange is that, mathematically, these three infinities are all the same size, even though the number of natural numbers must be the same as the number of odd numbers plus the number of even numbers.

Some infinities *are* bigger than others, however. In 1874, the mathematician Georg Cantor proved, for instance, that the number of 'real' numbers is bigger than the number of natural numbers. In other words, he showed that the infinity associated with all the integers and the decimals in between them is bigger than the infinity of just the integers. He went on to show that there are even bigger infinities than that — infinitely many of them, in fact. And then he went on to have a nervous breakdown.

If you're troubled by an infinity of infinities, you should take comfort in the fact that most of Cantor's contemporaries were neither willing nor able to grasp the idea either — it was this rejection of his work, not the concepts themselves, that led to Cantor's breakdown. But as we've said before, it's not natural to think like this. It takes ridiculous amounts of effort. If counting beyond three is unnatural, heading for more, towards an infinity of infinities that you can never actually comprehend, marks a commendable effort.

Assuming you're okay to carry on, we need to think about one more mind-bending thing before we tackle calculus itself: that we can work backwards towards infinity, too. As we have mentioned, as well as an infinite bigness, there is an infinite smallness. It's known as an infinitesimal.

Imagine slicing a cucumber into ever-smaller fractions. First you cut it in half, and then cut the half into two quarters. Take one of those and cut it into two eighths of the original cucumber. Now take one of the eighths and keep going. Eventually — in theory — you will have a slice so small that you can't describe it by any number, whether fraction or decimal. This is what we mean by an infinitesimal: almost, but not quite, zero. It's the closest you can get to nothing at all. The only thing smaller than an infinitesimal is zero itself. And when we slice time, distance or anything else into infinitesimals, we are in the realm of calculus.

The first person to attempt this was the German astronomer Johannes Kepler. He didn't do it to improve our understanding of the stars, however. He did it to save money on his wedding.[10]

In 1613, Kepler married for the second time. The wedding took place in Linz, Austria, and Kepler arranged for a wine merchant to supply a barrel of wine for the celebrations. However, he was appalled at the way the merchant worked out the cost of the barrel. First, he laid the barrel on its long side, with the bung hole at the top. He then inserted a stick into a hole at the top, and pushed it first down, then to the side until the stick hits the point where the long side meets the end face (the head). The price of the wine supplied depended on the length of the stick wetted within the barrel.

Kepler had already calculated the shapes of planetary orbits, described various optical phenomena, investigated the most efficient way to pack spheres, and shown that snowflakes have a hexagonal symmetry. He knew instinctively that this wine-pricing could be done better. Immediately, he pointed out that a long thin barrel could produce the same length of wetted stick while containing much less wine. Initially, he just argued with the wine merchant, but after the wedding he turned the dispute into a book, entitled *New Solid Geometry of Wine Barrels*, that he published in 1615. Within its pages, he uses a scheme that splits barrels into smaller and smaller circular slices in order to calculate the volume. He describes adding up the contributions to the volume from an infinite number of infinitesimally small slices.

Kepler also attempted to determine the optimal shape for a wine barrel — the proportions that would maximise the volume. He constructed a cubic equation that showed how the volume changed as the length of the barrel changed (for a fixed diameter) and determined that the greatest volume occurred at the turning point of this curve, which occurred when the length was $2/\sqrt{3}$ times the diameter. As it turned out, this was almost exactly the proportion used in Austrian barrel-making.

In this one endeavour we have the essence of calculus. First, it is about understanding the changes that occur in one quantity when another, related quantity is altered. That might be barrel volume with barrel dimensions, or the distance covered by a car whose speed is constantly growing after accelerating away from a standing start, or the number of people likely to be infected with a disease as its virulence increases over time. Second, it is about finding maxima and minima. What is the dose of a cancer drug that will give the most effective response? What quantity of fuel gives a 747 maximum range given that it burns fuel as it goes, but requires more fuel per mile travelled when it is heavier?

Kepler's wine-barrel work is generally considered as a precursor to calculus; a famous controversy still churns away over who *actually* invented calculus. That's because, in the second half of the 17th century, a long-running acrimonious dispute between Isaac Newton and the German polymath Gottfried Leibniz ended without a satisfactory resolution. However, the truth is that neither man did his work from scratch. Kepler's wine-barrel work aside, in the first half of the century, Pierre de Fermat had done some work on calculating the areas under curves, and finding the points at which curves reach a maximum or minimum. As we shall see, these are essential components of calculus, and Newton himself said that he arrived at the early form of calculus 'from Fermat's way of drawing tangents'. René Descartes would have turned in his grave at this; he had done similar work, and entered into a bitter dispute with Fermat over the issue of who had done it first. But an Italian mathematician called Bonaventura Cavalieri did work on infinitesimals that also set the stage for Leibniz and Newton, as did the English scholar John Wallis, who wrote it all down in a 1656 book called *The Arithmetic of Infinity*. In short, Leibniz and Newton are pre-eminent, but they built on the work of many others. So let's just stop arguing and get on with it, shall we?

Deriving solutions for HIV

Calculus is essentially an extension of algebra: it is a set of tools for finding out more about the lines and curves that algebra encodes. At school, we are not always taught this, though, and learn calculus as little more than a set of rules for abstract tasks. We learn to find the slope of the graph associated with a quadratic equation, for example, without ever really appreciating why we would want to do that. So let's start with one practical application: finding the gradient, or slope, of a curve that is associated with the way a fatal infection proceeds in the human body. This use of calculus, it turned out, was instrumental in protecting us from the worst ravages of the human immunodeficiency virus (HIV).

It's easy to forget just how bad things looked when HIV was at its most dangerous. After the first cases were reported in 1981, HIV quickly became the scourge of human populations across the globe. By 2007, HIV/AIDS had killed more than half a million Americans, and the US was refusing to admit any HIV-positive person into the country. In 2009, Washington, DC had a higher rate of HIV than West Africa — a 3 per cent prevalence — described by the District of Columbia Health Department as a 'severe and generalized epidemic'.[11]

Today, just over a decade later, HIV infection is no longer a death sentence. In fact, people with HIV live relatively normal lives. What happened? Calculus.

In 1989, Alan S, Perelson created a calculus-based mathematical model for HIV infection in the human body, and the subsequent fight between the virus and the human immune system.[12] He simplified the situation down to just four 'differential' equations that described what happens in the body when things like the concentration of the untreated virus in the blood change over time. Differential equations involve a process known as differentiation, which lies at the heart of calculus. It is, essentially, a process for finding the rate of change of something at one particular point.

Gradients change as you move up a curved slope

You could also think of differentiation as a way to quantify the amount of effort required to run up a slope. Running up the hill in the diagram above doesn't take the same amount of effort all the way up. The slope is steep at first, but then levels off. The hill is actually composed of a number of different gradients: they are large at first, but gradually get smaller as you progress up the hill. Differentiation is a way of finding the effort involved in running up that hill at a specific point on the hill's curve.

The process generally starts with the algebraic equation for the curve. The slope is calculated by finding the 'rise over run': the vertical change in y (dy is used to signify change in y) that occurs over a horizontal change in x (dx, the change in x). The slope is then defined as the rise divided by the run: dy/dx. This is sometimes known as the 'derivative' of the equation that describes the curve. The derivative is easy to find when dealing with straight lines. But what if we are dealing with a curve like the one below?

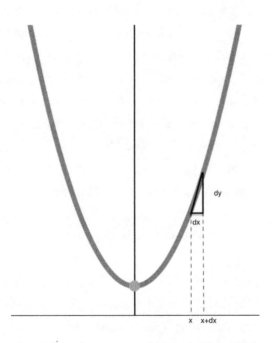

The gradient of a curve is rise (dy) divided by run (dx)

This curve can be described by the equation:

$$y = x^2$$

As we have seen, with curves, the slope changes as you move along them. In this situation, finding the slope at some point x is more of a challenge than with a straight line, where the slope is always the same. Because finding a slope means finding the rise and the run, that always involves two different points; the run is from one x value to another nearby one, and the rise is the change in the y value as we move between those two values of x. But on a curve, the slope will be slightly different at those two different points. So which value of the slope do we want?

Solving this problem involves a sleight of hand that makes the difference between the two points as small as possible — infinitesimally small, in fact. It's a little bit involved, but let's walk carefully through it to see where the general rule you learned in school comes from.

We'll follow through with $y = x^2$. We want to find the derivative — the slope — at some point x. In order to get a horizontal 'run' for the slope calculation, we'll run from x to a very nearby point $x + dx$. To be clear, dx is *tiny*. We can plug in the second value of x into the equation to give us $y + dy$, the second point on the rise between these two points. Because the curve is described by $y = x^2$ (in other words, x times x), $y + dy$ is given by $(x + dx)$ times $(x + dx)$.

To deal with this, we have to expand it out, multiplying each term in the first bracket by each term in the second bracket. That gives us

$$y + dy = x^2 + xdx + xdx + dx^2$$

Remember we said that dx was a tiny fraction of x? That means dx^2 is the square of a tiny fraction, which is an even tinier fraction. It's so small, in fact, that we can discard it. That leaves us with our second point:

$$y + dy = x^2 + 2xdx$$

To calculate the slope, we need the rise from y to $y + dy$. Our first point was $y = x^2$ and our second point was $y = x^2 + 2xdx$. So the rise dy, the difference between those two points, is $2xdx$.

The run goes from point x to point $x + dx$. The difference between those two points is dx. So the rise divided by the run is

$$\frac{dy}{dx} = \frac{2xdx}{dx}$$

The two dxs on the right-hand side of the equation cancel out (one is divided by the other, making 1, just as 3 divided by 3 is 1), leaving us with

$$\frac{dy}{dx} = 2x$$

To put it another way, the derivative of $y = x^2$ is $2x$.

You can follow the same process to find derivatives of other curves, but you end up spotting a general rule for how to do it: if

$$y = x^n$$

then

$$\frac{dy}{dx} = nx^{n-1}$$

Say I give you this equation for a curve:

$$y = 3x^2 + 5$$

To find the derivative I take the 'exponent' of x (the exponent is the raised 'power': 2, in this case) and multiply it by whatever number is in front of the x. If there is a term without x (here, there is +5), that just disappears. So in this case, the derivative is simply $6x$. So at the point corresponding to $x = 5$ on the horizontal axis, for example, the gradient of the curve is 30.

There are other rules about finding derivatives from other kinds of curves, and there are ways of dealing with combinations of these kinds of expressions. But essentially, it all boils down to finding the slope at a particular point.

And that's what Perelson did with his differential equations, which presented the rate of change of viral concentration of HIV, among other things, as the slope of a curve. His paper has differential equations for the T-cells, the macrophages, the virus, and the antigens, such as:

$$\frac{dV}{dt} = pI - cV$$

where I is the concentration of infected cells, p is the rate at which each infected cell produces new viral particles within the body, c is the rate at which the body's immune systems clears the virus, and V is the concentration of virus particles in the blood. dV/dt, as you might be able to guess by now, is the rate at which the concentration of the virus in the blood varies over time, a rate that corresponds to the slope of a curve showing a patient's progress over time. The full analysis showed researchers that there were different phases of the infection that could all be modelled mathematically.

In the year that Perelson's paper came out, the United States reached the milestone of 100,000 AIDS cases, and Congress established a National Commission on AIDS. Perelson's model was a lifeline. Within a short time, Perelson began working with clinicians and researchers to refine his model and the parameters within it. Perhaps the most significant partnership was with physicist-turned-biologist David D. Ho; together they used calculus to prove that it would take a combination of three 'antiretroviral' drugs to essentially rid the body of HIV.[13] That attack was 'triple therapy': a cocktail of three antiretroviral drugs that has turned HIV infection from a death sentence to a manageable problem.

There are countless examples of differential equations improving health care, from analysis of blood circulation to estimations of cancer spread and the effects of chemotherapy. But differential equations have changed the human experience in broader ways, too. If you drive or walk across a suspension bridge — the Brooklyn Bridge in New York, say, or the Akashi Kaikyō Bridge that crosses Japan's Akashi Strait — you're trusting an engineer's use of differential equations. It's not done just with maths, but that's where it starts, usually with a set of differential equations that describes the interplay between the mass, stiffness, and resistance to movement of the bridge's materials. They might, for instance, have to consider a differential equation that describes how changing the distance between the suspension wires affects their tension, and choose a configuration that also minimises the changes in tension as the load on the bridge increases (that is, the curve of tension against load would have a slope close to zero) to maximise safety. It's a similar story in a skyscraper: issues from how the load on the foundation changes as the height of the building rises, to how much it will twist and sway in a storm, are calculated using differential equations in the design phase. Any of the buildings, roadways, tunnels, and bridges in your environment that are less than 150 years old will have been designed using some form of differential calculus.

The Integration Game

Differentiation is only one side of the calculus coin. On the other side is 'integration'. This is actually the inverse of differentiation (although no one knew that when it was invented). Integration is about adding up the areas under tiny slices of a curve. Why would you want to do that? Because it's often the key to understanding how something — be it a nation's economy, a satellite in orbit, or a tropical storm — is behaving.

The textbook example tends to be a bit more mundane. Imagine you have a car that accelerates away from standing. After 5 seconds it is travelling at 36 miles per hour, and achieves a speed of 50 miles per hour after 10 seconds. If I draw a graph of speed against time, it will look something like the diagram below.

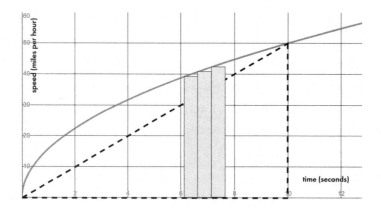

Methods for finding the area under a curve

Now imagine that you want to find out something new: how far this car has travelled in 10 seconds. All you have is speed and time. But consider this: speed is measured in units of miles per hour, and time is measured in hours (which we can split into seconds if necessary). If I multiply speed and time together, I am doing:

$$\frac{\text{miles}}{\text{hour}} \times \text{hours}$$

The result of that is just miles — a distance. In other words, multiplying the vertical measure by the horizontal measure, rather like finding the area of a square or rectangle, gives me some new information. The only problem is, this graph is not a square or a rectangle, so its area is not straightforward to work out. We could approximate the area using a right-angled triangle as shown with a dotted line, but it leaves out an awful lot of it. A better approximation comes from splitting the area under the curve into a number of rectangles (like the three grey ones) and adding their areas together. But even this is not terribly accurate — unless, that is, there are many, many rectangles, each with a negligible width. And by many, many I mean infinitely many. And by negligible, I mean infinitesimal.

Let's think about a situation where y depends on x in a way that traces out a smooth curve like the one above describing the speed of the car. Say we want to work out the area under that curve between the point where it crosses the y-axis, and the vertical line up from whatever value of x we're considering. Let's split that area it into thin rectangles. Each one has a width we'll call dx and — roughly — height y. So the area of each rectangle, which we'll call dA, is y times dx. So we have

$$dA = ydx$$

If we divide both sides by dx, we find:

$$\frac{dA}{dx} = y$$

Remember how we found dy/dx, the 'derivative' that describes how y changes as x changes? Now we've stumbled across a relationship between derivatives, area, and y, the points that describe the original curve. Integration comes out of that — it is, essentially, the 'opposite of differentiation'. Leibniz gave us the symbol for the integral, or sum: $\int$. This is what used to be called a 'long s', to reflect the fact that it represents an infinite number of tiny summations. It's usually written with a dx after the thing you're integrating, so that you know you're talking about the change as x changes. The integral of $y = ax^n$ is the area under that equation's curve, and can be written as:

$$\int y\, dx = \frac{a}{n+1}x^{n+1} + C$$

C is an unknown constant (if you remember, when we differentiate, we discard any terms that don't have x, so when we reverse the process, we have to bring them back in, albeit as unknowns).

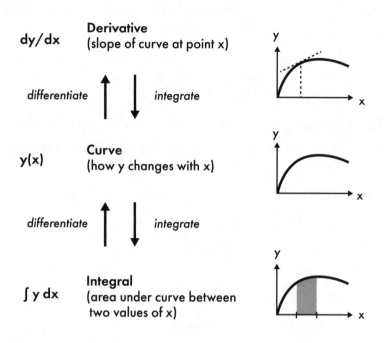

The relationship between curves and their derivatives and integrals

Like differential calculus, integral calculus is in play in myriad aspects of modern life. Weather forecasting and climate modelling involve integrating the amount of sunlight hitting the Earth's surface, for instance. Integrating predicted rainfall can determine whether there is a danger of flooding. NASA engineers use integral calculus to plan mission trajectories; when black pioneer mathematician Katherine Johnson calculated orbits for Alan Shepherd's 1961 *Freedom 7* mission, and for John Glenn's *Friendship 7* mission the following year, she had to use integral calculus. Luckily, she was a whiz at it — so much so that Glenn requested that she alone be trusted with checking the output from the new electronic computer.[14]

The Tortured Minds of Calculus

Integration is relatively straightforward these days, a part of the mathematician's standard toolkit. Nonetheless, it was incredibly difficult to discover. Leibniz and Newton came at it from different angles. Newton's derivation was so advanced for the time that he didn't really bother to explain it to his contemporaries. He once told them, rather arrogantly, that it would now take him 'less than half a quarter of an hour' to find the area under any curve. This astonishing speed, he added, was due to a 'fountain I draw it from, though I will not undertake to prove it to others'.[15] Instead, he relied on their knowledge of graphs and geometry to walk them through the basic principles involved.

No doubt Newton's contemporaries felt a little dumb. When we first come across it, calculus does that to all of us. But we shouldn't let it. Even though we can follow the steps to get the right answers, the ideas behind calculus, the derivations, and the leaps of imagination involved are truly staggering and extremely difficult. Archimedes took the first steps towards solving related problems more than a thousand years before Newton and Leibniz produced their solutions. Along the way, mathematical luminaries such as Descartes and Fermat only saw its shadow. To truly get to grips with it, you have to imagine infinities and infinitesimals. You have to recognise (as Fermat and Descartes didn't) that the tangent to a curve (the line that just touches it) tells you the slope of the curve at that point, and provides a way into understanding all the curve's properties. You have to be able to create long chains of sums that are known as infinite series, as both Newton and Leibniz did, and have the insight to see terms that will cancel each other out, exposing a hidden path through the dense undergrowth of your derivations. There's a reason why calculus was a millennium in the making.

You might, at this point, be asking yourself what kind of person can come up with all this. The answer is: not people you'd want to knock around with, for the most part. Pierre de Fermat, for instance, whose days were spent in the courtrooms of Toulouse as a lawyer and a judge, spent his evenings separate from his family, working on maths problems. He had no particular desire to share his discoveries with the wider world, publishing precisely none of the insights that he made. We only know about them because they were found in his notebooks and diaries after his death.

However, Fermat did write about some of those discoveries in letters to other mathematicians. In fact, this was how René Descartes heard about Fermat: through a mutual acquaintance called Marin Mersenne. Descartes was, unlike Fermat, rather full of himself. He was described by a contemporary as 'cold and selfish'. When he boasted that he had discovered a way to create tangents for any curve, he said that 'this is not only the most useful and most general problem in geometry that I know, but even that I desired to know'.

Descartes was devastated when he learned from Mersenne that Fermat had solved the same problem ten years earlier. To make himself look better, Descartes went through Fermat's proofs and publicly proclaimed that he had found a slew of embarrassing errors — which wasn't true at all.

Decades later, Newton paid Descartes back by describing him as one of the 'bunglers in mathematics'.[16] Newton may not have been a bungler, but he was, in general, extremely unpleasant. It is said that he rarely laughed, and he admitted that once, as a child, he threatened to burn down his mother and stepfather's house with them inside. He referred to those who tried and failed to understand his work as 'little smatterers', and actively avoided dealings with anyone he did not regard as his equal. 'I see not what there is desirable in public esteem, were I able to acquire and maintain it,' he once said. 'It would perhaps increase my acquaintance, the thing which I chiefly study to decline.'[17]

Leibniz, for his part, was also rather pleased with himself. His calculus work, he once boasted, consisted of 'the great part of the discoveries which have been made concerning the subject'. He was not a people person either — though at least he regretted this. He once moaned to a friend that he seemed to 'lack polished manners and thereby often spoil the first impression of my person'. He stayed single all his life, and had no children. Descartes and Newton also never married and left no descendants. At least Fermat had a family, even if he chose not to spend much time with them. None of these people, however, can hold a candle to the mean-spiritedness of the first people to really exploit Newton and Leibniz's work. Their lives became a soap opera of niggling acts of sabotage, gleeful gloating, character assassination, and good old-fashioned sibling rivalry. You've heard of the Borgias. Are you ready to meet the Bernoullis?

The Brawling Bernoulli Brothers

In the middle of the 17th century, the Bernoullis were known only as a Basel spice-trading family.[18] The mathematical bent arrived with Jacob Bernoulli, born in 1655. He followed his parent's wishes and studied theology at university, but supplemented his course with mathematics and astronomy — despite his parents displeasure at this unspiritual showing. Jacob toed the line enough to begin his working life as a protestant minister, but he had caught the maths bug: he studied it wherever he could, and began to teach mechanics courses at the University of Basel. Eventually, his religious calling faltered, but his academic research took off, and in 1687 he was appointed professor of mathematics. He studied Leibniz's calculus, working together with his younger brother, Johann.

Johann too was mathematically gifted. He had refused to embrace his father's desire that he take on the family spice business, and demanded that he be allowed to go to university. He was eventually permitted to enrol for a medical degree. However, he also soon began to study maths on the side.

The brothers' collaboration started off smoothly enough, and they did extraordinary work, making Leibniz's opaque and difficult calculus both accessible and applicable. With just a few days' study of one of Leibniz's texts on differential calculus, they worked out 'the whole secret', as Johann put it in his autobiography. But after a few years, the cracks began to show. From their writings, it is clear that Johann thought he was collaborating with Jacob, while Jacob considered Johann as nothing more than his student. Sibling rivalry soon began to colour their work.

In 1690, Jacob wrote a paper that used the word 'integral' to describe a method to find a cumulative property, such as the area under a curve. Johann would always claim afterwards that he had invented the term. He painted Jacob as a slow, plodding learner, and himself as a mercurial genius with near-revelatory insight. Then, in 1694, Jacob suggested in a published article that Johann's 'inverse tangent' method of solving a particular problem was inefficient and applicable only to a narrow range of examples. It was, he said, just a trick.

The gloves came off. Johann railed in a letter to a mutual acquaintance that his brother was 'filled with rage, hate, envy and jealousy against me. He holds grudges ... he finds it unbearable that I the younger brother receive as much esteem as he the older one does, and he would take great satisfaction in seeing me miserable and reduced to humility.' A while later he wrote to Leibniz himself, with whom the brothers had established a collaboration: '[Jacob] persecutes me arduously ... with his secret hatred.' Later still he accused his brother of being 'as secretive as Mr Newton'.

Meanwhile, Jacob complained to Leibniz of his brother's 'words filled with venom'. They entered into public duelling: with pomp and ceremony, Johann set Jacob a calculus-related problem carefully designed to bring his brother crashing down to Earth. Jacob returned the favour, setting a more difficult problem for Johann. An anonymous bystander to the procedure told Johann he would give him 50 silver ecus if he could solve it within three months. Johann offered up a solution that, he said, took him less than three minutes. But it was flawed. There was much back and forth, and Jacob openly mocked Johann's effort in a learned journal.

The watching community lost their fascination and became embarrassed. Their peers told the brothers they would be admitted to the Royal Academy of Science if they managed to resolve their differences. Jacob objected to the idea of Johann's admission to the Academy, and suggested to a friend that their peers had an 'excessively high opinion of the abilities of my brother'. By this point, Johann was a professor of mathematics at Groningen University. He was socially isolated — the family had sided with Jacob — but he was also defiant. 'I manage without him,' he wrote to a mutual friend. 'I do not depend on him in any way, and I owe him nothing.'

They were never reconciled. In 1705, Jacob fell so ill with gout that the family pressured Johann into travelling to see him. But Jacob died while Johann was still on the road. From beyond the grave, Jacob had the last word. Johann inherited Jacob's professorship at Basel, but nothing else: as he lay dying, Jacob had instructed his family not to let Johann have access to any of his writings, and they dutifully executed his wishes.

Johann seems to have taken out his frustrations on his son Daniel, who was just five years old when his uncle died. Daniel grew into a formidable mathematician who yearned to explore the powers of calculus. Inexplicably, however, Johann followed in his own father's footsteps and forbade Daniel from studying mathematics; he was permitted only to study medicine. Eventually, Johann deigned to teach his son some calculus, but was appalled when he showed an aptitude for the subject. In 1734, Daniel achieved joint first prize with his father in a competition set by the Paris Academy. Johann was so enraged at having to share the honour that he threw Daniel out of the house. A few years later, Daniel won the prize again, his father left far behind. With astounding spite, Johann announced that the winning work, published in a 1738 book titled *Hydrodynamica*, had been plagiarised from his own book *Hydraulica*. This, he said, brandishing the volume at the Academy, had been published in 1732. In fact, Johann had falsified the publication date of his work; it was published a year after Daniel's book, and plagiarised his son's work.

Showing enormous strength of character, Daniel tried on many occasions to reconcile their differences. He failed every time. But he didn't let that stop him from applying his father's (and his uncle's) mathematical innovations to major problems, and creating some of the most impactful applications of mathematics ever devised.

Differentiation, Disease, and Derivatives

'I simply wish that, in a matter which so closely concerns the wellbeing of the human race, no decision shall be made without all the knowledge which a little analysis and calculation can provide.' This is how Daniel Bernoulli introduced his 1760 paper suggesting that calculus be applied to the question of whether it was worth vaccinating a population against smallpox. His answer was an emphatic yes — and he had the data to prove it.[19]

According to Bernoulli's calculations, 75 per cent of the 18th-century human population had been infected with smallpox. The disease was responsible for 10 per cent of all deaths; in London alone the number of smallpox fatalities reached 15,000 in some years. Adults were, by and large, immune; if they had survived, it was because their bodies had manufactured antibodies to the disease. But children were vulnerable. So, should they be given the new vaccinations? Well, said Bernoulli, I have equations for that.

His first equation calculated the number of people who had never had smallpox, which was a fraction of the current population. His second equation forecast the annual number of infections and deaths due to smallpox, and the number of lives saved if smallpox were eradicated. Here is some of the original text:

the number of those who have not had smallpox at this age = s ...

the element $-ds$ is equal to the number who catch smallpox in the period dx, and according to our hypotheses this number is sdx/n;

for if, in the space of a year, 1 out of n catches smallpox, it follows that in a period dx there would be sdx/n who would catch the disease out of s persons ...

In his analysis we see that familiar notation of dx and ds — all due to Leibniz — and, later, talk of integrals and differentials. His conclusion was that the numbers were clear: France should vaccinate.

It was the first attempt to use mathematics to influence public health policy, and it could not have been done without calculus. Not that it worked; despite the maths, the citizens of France eschewed vaccination against smallpox.

Daniel Bernoulli's next realisation was that calculus could also be applied to economics. His first insight was the rather banal law of the 'diminishing marginal utility of money'.[20] In other words, if you have a lot of it, a small addition to the pile makes far less difference to you than the same addition would to someone with far less of it to start with. Or, as he put it, 'There is no doubt that a gain of one thousand ducats is more significant to a pauper than to a rich man though both gain the same amount.'

In calculus terms, x is your current wealth. Its utility to you is u. Bernoulli said that du/dx, the change in utility brought by an increase in wealth, diminishes as the wealth increases. It's hardly a great revelation. But it sowed the seeds of applying calculus to the investigations of economic theory. And this is a civilisation-changing genie that refuses to go back into the bottle.

Do you remember Thales of Miletus and his exploitation of the olive growers? Perhaps that was the point at which we should have realised that mathematics is power. According to Aristotle, Thales only wanted to make a point: that philosophers could easily make themselves rich but realised that there are more important things. What he may have unwittingly achieved, however, is the proof that, if getting rich is important to you, a grasp of mathematics is only going to help. No wonder Wall Street, the City of London, and every other financial hub across the globe swallow up calculus-savvy maths and physics graduates.

What started with Thales's foresight about the future value of oil presses has culminated in the attempted prediction of the future value of any commodity that might conceivably be used as a means of making money. As anyone who has ever had to make a decision about the stock market will tell you, financial trading is, essentially, gambling. Which is exactly why the mathematics behind finance has its roots in probability theory.

Probability theory began with Jerome Cardano, who was keen to make enough money at the card and dice tables to pay his way through medical school. But it was the calculus pioneer Pierre de Fermat, working with Blaise Pascal, who really got to grips with the subject.[21] Together, they analysed the odds of outcomes in various kinds of games, leading them to develop a formula that is more or less equivalent to something you might use to determine the value of the financial packages we call derivatives.

A derivative is essentially a contract between a buyer and a seller. They agree on the price at which some asset will be sold on some future date. Imagine you're trading in oil futures. You would enter into a contract to buy a specified number of barrels of oil at a specified price on or after a specified date. Your hope is that, by the time that date arrived, the price of oil would have risen beyond the agreed price so that you either make money on the trade, or can sell on the contract to a keen buyer before the date arrived. The problem is, you don't know exactly what the price of oil is going to do in the intervening time. And so you have to model the likely changes using mathematics.

The use of calculus in finance grew and grew after Daniel Bernoulli's initial insight. In 1781, the French mathematician Gaspard Monge used calculus to work out how to minimise transport costs when moving soil during the construction of forts and roads.[22] Monge's approach is, essentially, the same solution used in financial hedging, where people make investments that will minimise their overall loss should an unexpected problem hit some of their other financial activities. These days, calculus is applied across the financial markets, but one key equation stands out: the Black–Scholes–Merton model.

It all began in 1973, when Fischer Black and Myron Scholes, two academics working in economics, published a paper called 'The pricing of options and corporate liabilities'.[23] Soon after, economist Robert Merton published a development of the idea called 'On the pricing of corporate debt: the risk structure of interest rates'.[24] Though they might sound like a snoozefest to you (they do to me), these papers were so insightful, innovative, and influential that Merton and Scholes won the 1997 Nobel Prize for economics (Black had died of throat cancer in 1995, and was thus ineligible).

Black, Scholes, and Merton awakened an interest in something called an 'option contract'. This is similar to the oil futures we just mentioned: it is a contract between two people to buy and sell some commodity, or stocks, at a pre-agreed price within a certain date, should they want to execute the deal then. The buyer might, on the other hand, sell the option on to a third party. It's just another way to bet on the market value of a stock or commodity rising or falling.

The interest arose because Black, Scholes, and Merton had shown that you can use calculus to work out a mutually beneficial price for an option contract.[25] Specifically, they used a 'partial differential equation'. Where an 'ordinary' differential equation has just one changing variable (and is generally quite easy to solve), something with two or more changing factors is a 'partial' differential equation. An example might be the value of a stock that changes over time, but also changes with the value of a related stock: perhaps the price of oil, which varies depending on the amount released to the market, but might also depend on the price of gas. Partial differential equations often can't be solved properly at all — instead, they are solved 'numerically', which means using a computer to repeatedly try different combinations of numbers to see what works.

By making it possible to use calculus to put a value on things like options, Merton, Scholes, and Black changed the way money works in all the world's market economies. Their influence can be seen in the numbers. In 1973, at the time of publication, there were only 16 option contracts on the market. Now the market is worth trillions of dollars. Across the decades, researchers continued to innovate, creating new calculus-based ways to find value (and make money) in the financial markets. Most of these innovations involve solving partial differential equations in one form or another. And this is where the Black–Scholes–Merton model (and others like it) have led us into trouble.

Because of the complexity of these models, they were packaged up into computer programs that enabled traders to input a few variables related to current market conditions, and get an output that was, effectively, a recommendation for action. Unfortunately, few of these programs were explicit about their constraints. Black, Scholes, and Merton had been upfront about where and when the solutions of their partial differential equations would be valid and useful, but the small print on the new programs was frequently ignored — if it was supplied at all. And since no one using the program at the sharp end of financial transactions knew anything about the equations at its core, they weren't in a position to question its recommendations. The result was that more and more institutions accrued hidden, toxic debt.

The causes of the global financial crisis are enormously complex, but essentially it comes down to a lack of information about risk. Traders working for most of the big banking and finance institutions had unknowingly bought huge volumes of financial packages that concealed toxic debts. By the time it began to come out that firms such as Lehman Brothers owned debts that would never be repaid, there was nothing to be done about it: the firms could no longer trade. It was the Medici Bank all over again. In September 2008, Lehman Brothers went under. You know the rest.

It's not all doom and gloom, though. Daniel Bernoulli made a third contribution to the development and application of calculus. After health and finance, he applied Leibniz and Newton's equations to describe and predict the way fluids flow. And here we can rediscover the joy of calculus. The calculus of fluid flow is at the heart of the way we design aeroplanes. Get it right, and you can use it to win crucial victories that change the course of history — the Battle of Britain, for instance. Thanks to Daniel Bernoulli's work, we are ready to circle back to the Supermarine Spitfire; let's take to the skies, and soar on the wings of differential equations.

In Search of the Perfect Flight

Bernoulli began by examining the 2,000-year-old discoveries of Archimedes. They were relatively dull rules about the properties of fluids that sat, unmoving, in containers such as bathtubs. Bernoulli used calculus to spice things up, combining it with Newton's laws of motion. He published the results of his investigations in the book that his father would later plagiarise: *Hydrodynamica*.

One of Bernoulli's most significant discoveries was that if the speed of a flowing fluid increases, there is an associated decrease in the pressure that this fluid exerts on the world around it. Applied to aircraft wings, this can be used to explain the phenomenon of lift. When you use calculus to find the pressure variation over the wing's surface, you see an upward force.

The truth is, we don't actually understand what makes aeroplanes fly. Ridiculous as it seems, experts still argue over whether it is best to apply Bernoulli's principle, Newton's third law of motion — every action has an equal and opposite reaction — or some other explanation. But perhaps you would be swayed by the fact that the greatest scientist of the 20th century plumped for Bernoulli?

In 1916, having just published his general theory of relativity, Albert Einstein turned his attention to the question of flight.[26] He used calculus based on Bernoulli's principle to propose a new wing shape, where the top surface has a pronounced hump that would increase the speed of the air flowing over it, thus reducing the pressure in that region and causing the wing to experience a net upward force due to the air pressure below.

Einstein's wing performed poorly in wind tunnel tests, but his reputation meant that people were still willing to give it a go. The German aircraft manufacturer Luftverkehrsgesellschaft (LVG) built a prototype in 1917, and the flight pioneer Paul Ehrhardt volunteered to be the test pilot. The flight didn't go well. 'I hung in the air after take-off like a pregnant duck,' Ehrhardt later recalled.[27] Einstein was put off applied physics for life. 'I have to admit that I have often been ashamed of my folly of those days,' he once said of the experience.

Lesser-known scientists and mathematicians did much better, though. In fact, Einstein had — as was typical for him — ignored the astonishing progress that others had made. As we saw at the beginning of this chapter, at the time Einstein was playing around with the subject, other mathematicians were converting Frederick Lanchester's initial, instinctual parameters for wing design into mathematical equations also based on Bernoulli's work. And the 1920s saw a vast explosion in the scientific literature on flight, with many of the most significant discoveries made in Germany. That was where Beverley Shenstone spent two years working at the Junkers factory in Dessau. He returned to England in 1931, and a year later, Reginald Mitchell gave Shenstone a job at Supermarine on a wage of £500 per annum.

At that point, Shenstone didn't have all the calculus knowledge he needed to design the Spitfire. He clearly had some, though. The author Lance Cole went back through Shenstone's papers, books, and diaries to research his book *Secrets of the Spitfire*. On the inside back cover of a textbook on differential calculus that Shenstone had owned for nearly 20 years, Cole found 'a pencil-drawn, hand-created elliptical calculation ... with supporting calculus'. But Shenstone was aware of his need for more knowledge. We know that because when he published a paper on the calculus of wing design in 1934, he included a humble acknowledgement of a man whose contributions have almost disappeared from history: 'In conclusion the writer wishes to express his indebtedness to Professor R. C. J. Howland for valuable help and advice in preparing this paper.'[28]

Raymond Howland was a mathematician based at what was then called the University College of Southampton, on the south coast of England. Howland was a calculus specialist; when he and Shenstone met — by chance — the two talked about their work, and Howland became fascinated by Shenstone's attempts at applying the art. The result was a trade: Howland taught Shenstone advanced calculus, while Shenstone taught Howland aerodynamics.

Shenstone's public declaration of thanks to Howland came in the same year that Supermarine began their attempt to make a fighter with elliptical wings. 'The elliptical wing was decided upon quite early on,' Shenstone later wrote.[29] 'Aerodynamically it was the best for our purpose because the induced drag ... was lowest when this shape was used; the ellipse was an ideal shape, theoretically a perfection.'

In the end, the theoretically perfect shape had to bear some compromise. That December, Supermarine's craftsmen began construction of a prototype, and the final ellipse was 'simply the shape which allowed us the thinnest possible wing with sufficient room inside to carry the necessary structure and things we wanted to cram in,' Shenstone said. And, he added, 'it looked nice'.

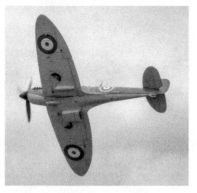

The elliptical planform of the Spitfire wing
Arpingstone, Public domain, via Wikimedia Commons

The wing planform — the shape as seen from above — had to be composed of a number of curves in order to meet the design specifications. All those curves had to be meshed together, with the gradients at each of their meeting points being identical in order to keep the overall shape smooth and aerodynamic. It's not clear just how much the final design owed to Shenstone's computations and how much was owed to the skill of Supermarine's 'lofters', the draughtsmen who traditionally worked in a mezzanine loft set above the factory floor. However, it's definitely a job you could do with calculus if you had been taught it well, and a paper that he and Howland published together in 1936 suggested that Shenstone had been taught very well indeed. It was called 'The inverse method for tapered and twisted wings', and described in complex calculus how changing the shape of wings affects their performance.[30]

Sadly, this paper marked the end of their collaboration. Howland died that year, never knowing the importance of his tutelage. The Spitfire was a triumph; some of the praise heaped on its handling seems almost as hyperbolic as the trajectory of a comet. Its pilots have called it a 'perfect flying machine' and 'quite out of this world'. But it is perhaps not surprising to hear how much British pilots enjoyed flying the Spitfire. What is surprising is how much the German pilots admired its manoeuvrability. As the Battle of Britain reached its most intense point, Field Marshal Hermann Göring asked what he could supply to the Jäger squadrons in France to help them overcome the British resistance. 'I should like an outfit of Spitfires,' said Gruppenkommandeur Adolf Galland.[31] Heinz Knocke, another pilot who flew against the Spitfires in the Battle of Britain, had similar feelings about the enemy's advantage. 'The bastards can make such infernally tight turns; there seems to be no way of nailing them,' he wrote in a book about his experiences.[32]

The Battle of Britain, Hitler's first significant military defeat, changed the outcome of the Second World War. Ralph Ingersoll reported that 'The *Luftwaffe* over England has never been the same since. Its morale in combat is definitely broken, and the RAF has been gaining in strength each week.' It made Hitler's ultimate defeat look possible for the first time, and encouraged the American people to join the conflict. That — along with establishing urban, financial, and medical miracles — is the power of calculus.

Students have always seen calculus as the watershed between standard and advanced mathematics. Somehow, everything we learn before calculus is relatively easy to absorb, and if calculus defeats you, chances are that you'll make little further progress. But if calculus was your breaking point, take heart. As we've seen, it took some of the best human minds to fully pin down the mathematics of change. Once we achieved that, though, we never looked back. Calculus became the multi-tool of mathematics, solving medical, military, financial, and architectural problems, to name just a few. The derivatives market, the Spitfire, triple therapy for HIV, and the Brooklyn Bridge are an impressive legacy for a discipline that started as nothing more than a playful dalliance with the mathematics of infinity.

The subject of our next chapter could hardly have a more different origin. One man, John Napier, invented the logarithm specifically to help astronomers with their sums. Like the calculus of the infinite, though, its applications seem to have found no limits.

Chapter 5

LOGARITHMS

How we launched science

The work of a Scottish laird, the logarithm is not much more than a tool for turning multiplication into addition, and division into subtraction. But that simplicity belies its pivotal role. It gave us error-free calculations of celestial orbits, and thus cemented the Sun's new place at the centre of our solar system. Once converted into a slew of mechanical calculation tools, it powered centuries of science and engineering, including the design and construction of the atomic bomb. It also introduced us to e, the mysterious never-ending number at the centre of myriad natural processes. And turned inside out, it describes the all-too-familiar exponential spread of infection at the outbreak of a viral pandemic.

In 1601, Johannes Kepler, the man who invented integral calculus to save money at his wedding, published his calculation of the orbit of Mars. It had taken him four years. Fifteen years later, he came across a new mathematical invention that would have saved him most of that time.

'A Scottish baron has started up, his name I cannot remember,' he wrote to a friend, 'but he has put forth some wonderful mode by which all necessity of multiplications and divisions [is] commuted to mere additions and subtractions.' It seems that Kepler was jumping for joy at the potential for future work saved — so much so that his mentor Michael Maestlin felt it necessary to tell him off. Kepler complained that his colleagues had told him 'it is not seemly for a professor of mathematics to be childishly pleased about any shortening of the calculations'.[1]

Try telling that to the incalculable number of people who, over the next 350 years, could not have done their jobs without that mathematical invention. It was called the logarithm and, as Kepler appreciated, it gave us a way of manipulating numbers to turn difficult calculations into easy ones. When transferred onto a set of wooden sticks known as a slide rule, logarithms powered centuries of science and engineering. The slide rule facilitated the Enlightenment, the Industrial Revolution, the nuclear age and the space race. If you want to get a sense of its importance and longevity, know this: Isaac Newton had a slide rule; the first steam engine was built with one; scientists used them at the detonation of the first atomic bomb; and the Apollo astronauts carried slide rules to the moon. The transport, industrial and housing infrastructure of the 20th century was designed using logarithmic slide rules — engineers literally wore them in holsters on their belts, like the essential tools they were. Logarithms could plausibly be cited as the single most influential invention in modern history. And they owe their existence to one man's extraordinary tenacity.

The name that Kepler could not remember was John Napier. Born in Edinburgh in 1550, John Napier was a committed Protestant who lived at a time of virulent sectarianism — and he was certainly infected. Nobly born to the title of 8th Laird of Merchiston, Napier was consumed by hatred for Catholics. Even in mathematics, his zealous personality shone through. Napier's primary passion was to uncover hidden knowledge through interpretation of the Bible's numbers. First, he took on the task of predicting the date on which the world would end. He attempted this through an analysis of the Revelation of St John, but could not come to a definite conclusion and settled on an upper limit of 1786. Any increase in collective human sinfulness would most likely, he said, bring the date forward. His second mathematical crusade set out to prove that the sitting Pope and the Antichrist were one and the same. It took a lot of effort, and some cavalier twisting of the scriptures, but Napier's knack for numbers meant that he got there in the end. The resulting treatise, *Plaine Discovery*, remained Napier's proudest creation until his death, he said.[2]

What Kepler called Napier's 'wonderful invention' had nothing to do with religion, though. The logarithm — the name is a conjunction of the Greek *logos* (proportion) and *arithmos* (number) — arose from a bout of pity for astronomers.

Anyone who wanted to chart the course of heavenly bodies, and create celestial maps useful to astrologers, astronomers, or sailors, had to fill reams of paper with calculations based on trigonometry. Angles measured with a sextant, and their sines and cosines, allowed the changing positions of the stars and planets to be mapped and predicted. But those calculations involved the multiplying, dividing, squaring, and cubing of great hosts of numbers. Napier realised that it was an appalling waste of everyone's time for these manipulations to be repeated anew every time anyone wanted to make an observation. And even setting aside the problem of what he called 'the tedious expense of time', there was the problem that people routinely made mistakes. 'I began therefore to consider in my mind by what certain and ready art I might remove those hindrances,' he says in the preface to his 1614 book that offered the solution. The book was published with the rather bold title *Mirifici logarithmorum canonis descriptio*, or *A Description of the Wonderful Law of Logarithms*.

Napier describes the contents as 'some excellent brief rules' for saving time on multiplications. The rules might have been brief, but their derivation was not. It had taken Napier two whole decades to create the book's 10 million entries. But the work had been worth it. Kepler, for one, was so grateful that he dedicated his 1620 publication *Ephemerides* to Napier, unaware that the originator of the logarithm was already dead.

Exponential Growth

As I write this, in March 2020, the news is full of the kind of maths that Napier pioneered. Logarithms are only being mentioned occasionally, though; mostly, it's mentions of their inverse: 'exponentials'. These are what define the rise in global cases of infection with Covid-19. When you plot out the increase in cases over time, you get a quickly rising 'exponential curve', the kind of curve we were all encouraged to 'flatten' by adopting infection-halting behaviours such as wearing face masks and social distancing.

'Exponential' is not an uncommon word, even outside of a viral pandemic. We use it casually to mean something that grows really, crazily fast. But, strangely, we have no real intuition for the exponential scale. When we're told things are growing exponentially, and we're asked to predict what state they might be in at a later time, we vastly underestimate the growth. That's because our brains shy away from extremes, normalising the growth in our imagination to make it more or less linear.

In a talk he gave thousands of times, the American physicist Allan Albert Bartlett presented a beautiful example of the mind-addling nature of exponential growth.[3] Imagine it's 11.00 am, he said, and he's handing you a bottle that contains one bacterium. He tells you that the bacterium's natural replication by division means the number of bacteria in the bottle will double every minute, and that the bottle will be brim-full of bacteria in 1 hour.

That seems, perhaps, just about believable. But, Bartlett asks, how full is the bottle at 11.56, just 4 minutes before the end? The maths tells us it is only 3 per cent full. That's rather counterintuitive: at that time, if you were a bacterium in the bottle, you'd have no sense that space is going to run out so soon. Even with 2 minutes to go, the bottle is only one-quarter full. It's half full at 1 minute before noon. And then, with the last doubling (which takes place in the last minute) the bottle is full.

Even more astonishing is what happens if Bartlett gives you three new bottles at 11.58, when the first bottle is only a quarter full. You look at the mostly empty bottle in your hand, and the three more on the shelf. It looks like you have plenty of time before all this space is used up. But you're wrong. The second bottle is full at 12.01 pm. By 12.02 all four bottles are full.

After 58 minutes of the bacterial growth you were optimistic. Four minutes later, your optimism is utterly confounded. This, Bartlett said, is the tragic story of our understanding of exponentials, whether it's applied to outbreaks of disease or to the unsustainable growth of human populations. At the beginning of his talk, he puts it like this: 'The greatest shortcoming of the human race is our inability to understand the exponential function.'

This shortcoming has a name. It is exponential-growth bias (EGB), and it applies beyond the world of epidemics and population growth. In fact, much of the research literature on EGB focuses on its relevance to personal finance — compound interest, in particular.

Compound interest is the kind that makes a prudent saver more money. Instead of just getting interest on the initial amount invested, the saver puts the interest earned — after a year, say — into the account. Then this can earn interest too. If you save £100 at 10 per cent annual interest, you'll have £110 at the end of the first year. But at the end of the second year, you'll earn 10 percent of £110, which is £11. So at the end of the next year, you'll be getting 10 per cent interest on £121. With compound interest the return on investment keeps growing — exponentially.

Unfortunately, the same also applies to loans. If the amount of debt you rack up on a loan grows exponentially because of non-payment and interest charged, this can lead to significant holes in your finances.[4] Even worse is the fact that we don't just underestimate the rate at which exponentials grow; we also *overestimate* our ability to deal with the numbers. In other words, we get exponentials wrong, and we do it with a dangerously misplaced confidence — which means we tend not to check our intuition or ask finance professionals for help.[5]

Applied to epidemics, EGB gives us a similar false sense of security.[6] At the beginning of a viral outbreak, the number of new infections per day tends to rise exponentially, as you can see on the illustration below of the way COVID-19 infections grew in the US during March 2020. But we look at the initial numbers and our brains tell us that the growth is only linear.

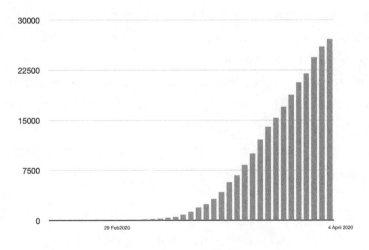

Daily new US cases of COVID-19 in early 2020 (source: CDC)

Say there were 50 cases on day 1, and 100 cases on day 2. Exponential-growth bias means we naturally assume there will be another 50 cases on day 3. But if the growth is exponential, going from 50 to 100 in a day means there's a daily doubling. So day 3 will see 200 cases, not 150. By day 10, there are actually over 25,000 cases more than in our assumed linear scenario. This leads to complacency: we assume that we'll come into contact with far fewer infected people than we actually will. Sometimes, our unmathematical brains are more dangerous than we know.

The Leap to Logarithms

The word 'exponential' comes from 'exponent'. You'll have seen exponents: they are the little, raised numbers that tell you how many times to multiply the number they refer to by itself. So when we say $2^3 = 8$, what we're actually saying is 'multiply the number 2 by itself three times'. That is, $2 \times 2 \times 2$, which makes 8. But we can also turn this relationship between the numbers on its head. This is where the logarithm comes in. Instead of focusing on the exponent, we can express things differently: we can say 'the logarithm in base 2 of 8 equals 3'. That might not sound to you like it's worth doing, but fortunately John Napier knew better. He realised that it provides a way to turn the onerous task of multiplication into the easy one of addition.

Mathematicians have long understood the relative difficulty of these two procedures. There is a wonderful (and possibly apocryphal) story of a German merchant of the 15th century, who wants his son to be educated in mathematics.[7] The merchant asks a local university professor for advice.

'If you want him simply to know how to add and subtract,' the professor says, 'a German university education will suffice.'
'And if I want him to be able to multiply and divide?' the merchant asks.
'Oh, then you must send him to Italy.'

The obvious implication is that the Germans weren't yet sophisticated enough to do multiplication. But Napier showed there was no need to go to Italy, where the Antichrist had his throne. Instead, he proved that you can get multiplication from addition via trigonometry.

Do you remember how trigonometry formed its sines and cosines? It was by looking at the lengths of two sides of a triangle inside a circle of radius 1. It turns out that there's an interesting by-product of this. Take two angles, A and B, and find their cosines. Multiply them together, and then double the result. That number turns out to be equal to the cosine of $A + B$ added to the cosine of $A - B$. In mathematical language:

$$2\cos(A)\cos(B) = \cos(A + B) + \cos(A - B)$$

That means you can use trigonometric tables to find the result of multiplying two numbers together. If you want to multiply X and Y, set X equal to $\cos(A)$ and Y equal to $\cos(B)$. Open up the booklet of trigonometric values and find A and B. Then work out $A + B$, and $A - B$. Go back to your trig table and find the cosines of these results. Add those things together, and you've got 2 times X times Y. Halve that, and there's your answer.

You can apply this process to any numbers you want to multiply, as long as you've got a set of trig tables. Napier was aware of this technique, and others — you can perform similar tricks with sines of angles, for instance — and knew that sailors and astronomers routinely used these 'trigonometric identities' to calculate their way around the heavens. But he became particularly interested when an acquaintance called John Craig told him that Tycho Brahe had used these techniques to achieve a great discovery.[8] Craig had seen Brahe and his assistants working with the technique when he visited Brahe's observatory on Hven island, where he stayed in Brahe's Uranienborg (literally, 'castle in the sky'). Brahe had used the trigonometric identities during the process of discovering a new star, and Napier clearly felt further discoveries might be accelerated if more astronomers could use them easily — especially if all the hard work was done for them. And so he set out to do the hard work himself.

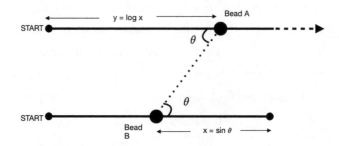

John Napier's method for calculating logarithms

Napier kicked off the logarithm revolution by imagining two beads moving along parallel strings, one finite and one infinite. The top string has infinite length, and bead A moves along it at a constant speed. The numbers defining its position grow in an 'arithmetic' progression, which means they increase steadily, with the same amount added at each step. After 1 second, say, you are at position 100; after 2 seconds, you're at 200; after 3 seconds, you're at 300. The bottom string is finite. Bead B starts level with the top bead, and starts at the same speed, but its speed decreases as it moves along. Specifically, the speed is proportional to the distance to the end of its string. It might start at speed 100, say, and then after 1 second its speed is 50; after 2 seconds its speed is 25, and so on. That has two consequences. First, it means that the numbers defining the lower bead's position decrease in a 'geometric' progression, which means they correspond to multiplication by, not addition of, a number. Second, it means that, at any moment after they start moving, the top bead will always be further along than the bottom one. And if you draw a line connecting them, the angle between that line and the bottom string gets smaller and smaller.

This is actually a way of providing a seemingly endless supply of triangles. As you can see, this connecting line is the hypotenuse of a right-angled triangle. The cosine of the ever-shrinking angle depends upon the distance the upper bead is ahead. Napier first defined the sine of the angle as the distance not yet covered on the bottom string. He then defined the number he was actually interested in: the logarithm of this sine. This, he said, was the distance the upper bead had travelled along the top string at this moment. For every possible angle, in increments of 1 minute (a minute is one-sixtieth of a degree — a Babylonian legacy measure), he computed its sine, the length remaining on the lower string, and the length covered on the top string, the logarithm. In order to achieve the kind of accuracy that would be useful to astronomers and navigators, Napier went to extraordinary lengths in his work. He set the length of the lower string to 10 million units, which allowed for seven decimal places. He wanted the logarithm to increase from zero by 1 each step. This resulted in the eye-watering 10 million values he entered in his logarithm tables, each one calculated via a painstaking and utterly exacting mathematical procedure. Now an astronomer could take cumbersome calculations involving multiplication and division, and convert them into calculations that involved just adding and subtracting some related numbers that they could look up in Napier's tables. This was one man's work, undertaken for twenty years, only to ease the work of others. Has there ever been a more selfless act?

A Change of Base

Napier must have breathed a huge sigh of relief when he was finally ready to publish his tables. But, as it turned out, the work was far from over. A London-based mathematics professor called Henry Briggs read Napier's book, and was hugely impressed. 'I never saw a Book which pleased me better, or made me more wonder,' Briggs wrote to his friend James Ussher.[9] But, Briggs added, it still needed a few tweaks.

As people, Napier and Briggs were chalk and cheese. Briggs was the Gresham Professor of Geometry, a no-nonsense Yorkshireman with few, if any, leanings towards fanatical religion, mysticism, or spirituality. Napier, as well as being a staunch Protestant, fancied himself as something of a magician. He practised astrology, and there are hints that he practised darker arts too. In 1594, he entered into a contract with a baron called Robert Logan, who engaged Napier to use whatever means he felt appropriate to find treasure lost somewhere within Logan's Fast Castle fortress. Napier's twenty years of seclusion led to him being labelled (long after his death) a suspected Satanist in the 1795 *Statistical Account*, a set of parish reports written by Church of Scotland ministers. 'It was formerly believed and currently reported that he was in compact with the devil; and that the time he spent in study was spent in learning the black art and holding conversation with Old Nick,' the *Statistical Account* says.[10]

No matter: Briggs was Napier's biggest fan. They exchanged letters, and Briggs planned a trip to Edinburgh. 'I hope to see him this Summer if it please God,' Briggs wrote to Ussher in 1615. And see him he did. In fact, according to one contemporary source, they stared at each other in mutual admiration for a quarter of an hour before either of them spoke.

Eventually, though, they got down to business. Briggs's suggestion was that Napier's logarithms were fine for performing trigonometric calculations, but should be tweaked so they were easy to use for ordinary numbers. Napier had chosen 10 million as a useful figure that would allow him plenty of decimal places to work with. But, as Briggs pointed out, that left an over-complicated situation.

Briggs saw straight away that Napier's set-up meant he had inadvertently created the situation where

$$\log (A \times B) = \log A + \log B - \log 1$$

Because of the way Napier had designed his logarithms, log 1 was not equal to zero. It was Briggs' suggestion to alter the basis of how the logarithms were calculated in a way that makes log 1 equal to zero. Then we are left with the rather desirable situation where

$$\log (A \times B) = \log A + \log B$$

This gives an unbelievably neat and tidy way of linking addition and multiplication.

At heart, logarithms are just a way of expressing a relationship between numbers. As we saw, $2^3 = 8$ expresses the same information as 'the logarithm in base 2 of 8 equals 3'. But with logarithms, we can change the 'base' to make calculations easier. One of the most useful bases, as Briggs realised, is base 10, where the logarithms deal with powers of 10 quite easily. Because log 1 was defined as zero, log 10 would be 1, log 100 would be 2; log 1,000 would be 3, and so on. You can see that the logarithm describes how many zeros come after the 1 in the Hindu-Arabic notation. Since 100 is 10×10 (ten squared, or 10^2); 1,000 is $10 \times 10 \times 10$ (ten cubed, or 10^3) and so on, it becomes clear that this improved logarithm is intimately and very simply related to the processes of multiplication.

It was clear to Briggs — and eventually to Napier — that the complex calculations of astronomers and other users of Napier's system would become much, much easier. All the pair had to do, then, was to recalculate the 10 million entries in Napier's book of logarithms. And that's exactly what they spent a good portion of the next two years working towards.

Their collaboration ended when Napier died of gout in the spring of 1617. But Briggs pressed on. The tables were completed (with the help of a Dutch mathematician called Adriaan Vlacq) and published in the Dutch town of Gouda in the summer of 1628. These were the 'base-10' logarithms we know today, using the natural numbers from 1 to 100,000 computed to 14 decimal places. The publication also gave tables of natural sine functions to 15 decimal places, among other trigonometrical data. Two years after publication, Briggs followed Napier into the grave, but the pair's legacy was complete.

Calculating Made Easy

Pierre Simon Laplace remarked later that the work saved by logarithms must have 'doubled the life of the astronomer'.[11] Kepler's life wasn't just doubled, though; the invention of logarithms seems to have affected the very way he thought. There are reasons to believe that Kepler's third law of planetary motion — among the most revolutionary insights of scientific history — owes an enormous debt to the discovery of these number proportions.

Kepler published his first two laws in 1609, but only hit upon the third in 1618, two years after he first saw Napier's work. The third law mathematically relates the time taken for a planet to orbit the Sun to the spatial length of the longer, 'major' axis of its orbit. In mathematical terms, the square of the orbital period is proportional to the cube of the 'semimajor axis' (the semimajor axis is half of the major axis). Kepler hit upon this not in terms of cubes and squares, but in terms of ratios. He said, on 8 March 1618, that it 'appeared in my head' that 'the proportion between the periodic times of any two planets is precisely one and a half times the proportion of the mean distances.'[12] We can translate that in terms of logarithms of the ratios of the periods and the distances. If planet A takes T_A to orbit the Sun, and has orbital radius (its average distance from the sun) r_A; and planet B takes T_B to orbit the Sun and has orbital radius r_B, then

$$\log\left(\frac{T_A}{T_B}\right) = 1.5 \log\left(\frac{r_A}{r_B}\right)$$

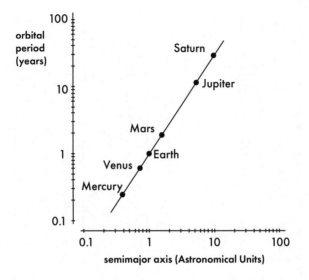

Johannes Kepler saw a logarithmic relationship between the planets' orbit size and the time it takes them to go once round the Sun

If we plot this out in what we know today as a log–log graph, the relation seems obvious. At some point between 1609 and 1618, something in Kepler's brain appears to have made a logarithmic leap. It only seems fair to suggest that Napier (and Briggs) may have made a huge, unanticipated and almost entirely unappreciated contribution to astronomy.

Even more work was saved by the automation of these calculations. Napier's first venture in this direction was with some wooden sticks that came to be known as Napier's rods (or Napier's bones, when later editions were made from ivory). As with his logarithm, Napier's rods were designed to turn difficult calculations into easy ones. The rods were divided into squares, each of which was divided diagonally into two triangles. In each triangle a number was inscribed, and the arrangement of the numbers turned these sticks into calculating tools that require addition skills but not multiplication.

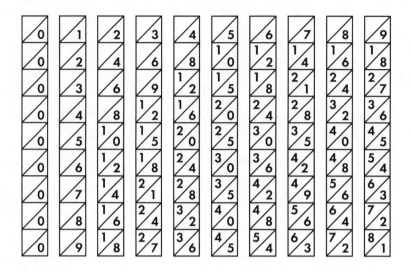

A set of Napier's rods

Say you want to multiply 423 by 67. You would take out the rods with 4, 2 and 3 at the top and place them side by side. Now write out the figures from the sixth row, adding together any diagonals: you have 2, 4+1, 2+1 and 8, so the number is 2,538. Now do the same with the seventh row: you have 2, 8+1, 4+2 and 1. That gives 2,961.

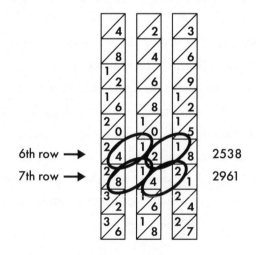

Using Napier's rods to multiply 423 by 67

Now you add these two four-digit numbers, but with the one from the sixth row shifted one place to the left (because the 6 in the original sum represents 10s, and the 7 is just units). So you're adding 25,380 to 2,961. That gives you 28,341, which is indeed what you get when you multiply 423 by 67.

Since they made multiplication, division, and finding square roots easier, Napier's bones became an extremely popular tool, and were produced in a variety of forms — including ones that calculated square roots and cube roots — before being developed into more complicated instruments. Some were automated: a machine produced by Wilhelm Schnikard in 1623 even added the relevant numbers together for you so you didn't have to do that in your head. In the 1650s, Pierre Petit, a French engineer, mounted the numbers on paper strips that he looped onto a drum so that they could all move relative to each other and make the process of multiplication even easier. Soon afterwards, the German polymath Athanasius Kircher went one better and crafted a multiplication machine that used Napier's rods, plus additions of the inventor's own, to perform the required calculations at the turn of a handle. But however successfully such machines automated multiplication, their impact was nothing compared to that of the semi-automatic multi-purpose mathematical weapon we know as the slide rule.

The Calculator Behind Centuries of Progress

One of the first consequences of Napier and Briggs's tables of logarithms was the realisation that tables are not strictly necessary. Instead, you can write out numbers on a logarithmic scale, where there is the same amount of space for the gap between 1 and 2 as between 2 and 4, or between 4 and 8.

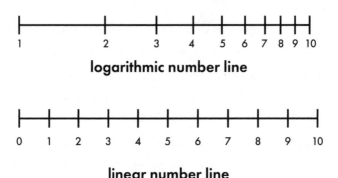

logarithmic number line

linear number line

The logarithmic number scale

Two of these scales running against each other provide a way of performing your calculation; the two scales become, effectively, a portable table of logarithms. The first person to do this was another Gresham professor: Edmund Gunter. He engraved the necessary numbers on a two-foot-long piece of wood that came to be known as a Gunter's scale. Using a pair of mathematician's dividers — a jointed pair of sticks whose sharpened ends can be set to measure the distance between two points — then allowed a mathematician to calculate sums and differences just by length measurements. Gunter combined his logarithmic scale with markings that denoted rhumb lines, return courses, and trigonometric functions, giving sailors a multi-functional tool for navigation that was in use for hundreds of years.

Landlubbers soon had an even better tool, thanks to William Oughtred. His innovation was to combine two pieces of wood, each marked with numbers on a logarithmic scale, in an arrangement that allowed them to slide along each other's length. The way the scales were arranged enabled the suitably trained user to perform all kinds of calculations. Oughtred's 'slide rule' was revolutionary for anyone who wanted to perform quick, accurate calculations. Isaac Newton was clearly a fan: in 1672 he explained to a John Collins how he used one to solve a cubic equation. Collins was interested because the same technique could be used to find the volume of liquid in a partially filled barrel — yet another application of mathematics to taxation.[13]

Newton also saw how to improve the design of the instrument: he was the first to suggest a moveable cursor, now a feature of almost any slide rule you'll see. At the end of the 18th century, James Watt developed a further improvement, with scales suited to engineering calculations, which he called the 'Soho'. He used it to calculate the necessary specifications of his newfangled steam engine, and many of his contemporaries took up the Soho slide rule as an aid to their work. It's clear from Watt's work that the Industrial Revolution was launched on the back of logarithmic technology. Chemist Joseph Priestley used one to process the results of his experiments and determine the chemical composition of air. And so important had the slide rule become that, across the English Channel, the qualifying exams for French civil servants required that they demonstrate a proficiency in its use.[14]

Demand for the logarithmic slide rule hit its peak in the 20th century. Science, engineering, and industry were thriving. They were also mathematically demanding: a slide rule was an essential tool in laboratories, on factory floors, and in design workshops across the Western world. And the technology kept evolving, offering more functions and better accuracy. In the first decade of the 20th century alone, 90 different new designs were registered. Nobel Prize–winner Julius Axelrod used one in work that led to the development of modern antidepressant medications called selective serotonin reuptake inhibitors. Katherine Johnson used one when she calculated the trajectory for Alan Shephard's journey as the first American in space. Johnson also used one to calculate the trajectory for the Apollo missions to the Moon. At the same time, NASA's engineers were using slide rules — they called them slipsticks — to design and build the rockets and landers for the Apollo missions. In fact, a slide rule was standard issue for the Apollo astronauts; Buzz Aldrin had to have one to perform last-minute calculations for the 1969 lunar landing. In 1969, his Pickett Model N600-ES (Eye Saver) Log Log Speed Rule would have sold for

$10.95. In 2007, it was put up for auction, and someone bought it for $77,675.

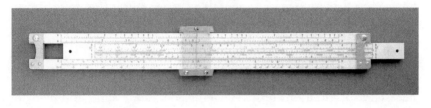

Buzz Aldrin's slide rule
Heritage Auctions, HA.com

Impressive as that figure is, there is a slide rule that played an even more significant role in human history. It belonged to physicist Enrico Fermi, and it helped him create the world-changing technology of the atomic bomb.

Birthing the Bomb

Witnesses report that, at 3.25 pm on 2 December 1942 a smile broke across Fermi's face. He was standing among a group of scientists and engineers in a squash court beneath the West Stands of Stagg Field at the University of Chicago. Apparently, he closed his slide rule and turned to his colleagues. 'The reaction is self-sustaining,' he announced. 'The curve is exponential.'[15]

The reaction in question was the first controlled nuclear chain reaction. Fermi had spent years experimenting and calculating, slide rule in hand, to see whether it was possible to create such an event — very much an open question at the time, and now, of course, appreciated as a hugely important one. This is the reaction that led to the creation of the atomic bomb, and to the development of nuclear power. This moment, created by logarithmic slide rules and the exhaustive study of exponential curves, set the global narrative for the next half-century.

Born in Rome, Fermi had come to Chicago via Stockholm. In 1938 he had been awarded the Nobel Prize for physics 'for his demonstrations of the existence of new radioactive elements produced by neutron irradiation, and for his related discovery of nuclear reactions brought about by slow neutrons'. The neutrons were released from the nuclei at the heart of atoms of beryllium, uranium, and other radioactive elements. Fermi had learned to slow the neutrons down — initially using nothing more technological than a block of paraffin wax — so that they would interact with the atoms in other metals and release nuclear particles that would make those metals radioactive too. But the ultimate aim was to make that process self-sustaining, using neutrons to release more neutrons that would, in turn, release more neutrons. The process behind that release unleashed vast amounts of energy. If it could be controlled, the world would have a new and almost unlimited source of power.

At the Nobel Prize ceremony in Stockholm, Fermi wore the customary black tie and tails. When the photographs of the ceremony were published in the Italian press the next day, there was an uproar. Commentators said he should have supported his government, the dictatorship led by Benito Mussolini, by wearing a fascist uniform to the ceremony and giving a fascist salute to the king of Sweden. Incensed by the fascists, Fermi did not return to Italy. Instead, he went to America, where he settled in New York to work on his ideas alongside the physicists at Columbia University.

Once the idea of a sustained, significant energy release started to become a possibility, one of Fermi's Columbia colleagues, Leo Szilard, composed a letter to President Roosevelt. It suggested beginning an atomic energy programme, and warned of the dangers if such a technology were to fall into the wrong hands. Albert Einstein delivered the letter, and the Manhattan Project was born.

Fermi knew that an atomic bomb could not be built if each speeding neutron only released one neutron from its target, or if there were insufficient neutrons in the target. That meant he had to perform intricate calculations of the radioactive emissions from each atomic nucleus, and of the 'critical mass' — the minimum amount of radioactive material required for a self-sustaining reaction. Those calculations, always performed with a slide rule, showed that if 100 speeding neutrons released just 103 or 104 other neutrons — slightly more than one each — the reaction would be self-sustaining, growing exponentially. Uranium could do this: Fermi's calculations suggested that splitting a uranium nucleus should release 1.73 neutrons, on average. And so he designed an experiment. His first pile of seven tons of uranium oxide cubes (and graphite blocks to control the reaction) was 11 feet high. The Columbia football squad was called in to lug the heavy blocks into place.

Fermi referred to the tower as his 'exponential pile' but the reaction, once started, was not a runaway exponential. The practicalities of the set-up meant each neutron was producing an average of 0.87 other neutrons. Improvements in the design took that number up to 0.918, but it was still not enough. The team needed a bigger, redesigned pile, and they decamped to Chicago to build it. It was potentially so powerful that Fermi designed a neutron-absorbing cadmium 'zip' rod that could be dropped into the pile in order to avoid a dangerous, over-steep exponential production of neutrons. On the day the Chicago pile went exponential, he used his slide rule to calculate exactly how much of the rod should be inserted to keep the observers safe.

This was not the only time Fermi's slide rule made a cameo appearance at a pivotal moment in history. He also had it with him when the US exploded the first atomic bomb on 16 July 1945. Fermi was in the observation bunker 20 miles northwest of the Trinity test site, and used the motion of scraps of paper to measure the force of the shock wave. When the papers blew away, Fermi took out his slide rule, performed a few calculations, and declared that the explosion had been equivalent to 10,000 tons of TNT.[16] He was wrong, as it turned out — it was more than 20,000 tons. But a slide rule, like any tool, can only work with what it's given.

Further developments of the slide rule remained in use well into the 1970s. Even if you have never used one, it may well be that your parents or grandparents have a slide rule in the house somewhere. You might even have a watch that boasts a circular slide rule on its rotating bezel. Pilots can often be seen with them: many are still trained to perform calculations of speed, distance, altitude, and fuel consumption in the old-fashioned way. However, almost as soon as the electronic calculator was invented, the slide rule became obsolete, and we quickly forgot that logarithms built our world.

The Wonder of *e*

William Oughtred certainly wasn't to know his logarithmic slide rule would have such an important role to play in the human story. But he doesn't seem to have been the kind of man who worried much about his place in history. There is an appendix in the second edition of Napier's first book of logarithms, published in 1618, that scholars believe was written by Oughtred.[17] It's frustrating that he didn't sign it because it is another historic moment: our first glimpse of an extraordinary number we now call '*e*'.

Like Briggs, Oughtred seemed to have noticed that Napier's choices concerning the calculations of his logarithms could be improved upon. His appendix (if it is his) contains a table of logarithms that has nothing to do with the ones that Napier constructed. Next to 8, for instance, is the number 2079441. These are the digits (minus a decimal point after the 2) of what we now call the 'natural logarithm' of 8. In other words, 2.718281828 raised to the power 2.079441 is equal to 8.

You might reasonably ask, what on earth is 'natural' about any of those numbers? Well, when it comes to 8 and 2.079441, nothing. But Oughtred's choice of 'base', that 2.718281828 — and that's not the end of it; it goes on and on forever — seems to lie at the heart of myriad natural phenomena.

No one knows why Oughtred put this table into the book. There is no explanation, just an implicit assumption that it will be useful to readers. In fact, natural logarithms don't pop up again for another 65 years, when Jacob Bernoulli made a calculation about the nature of accruing interest.

In 1863, Bernoulli was working on a problem related to how often you'd like a bank to add the interest to your account. Imagine you had $1,000 and (in this highly fanciful thought experiment) they were giving you 100 per cent interest yearly. If they added it at the end of the year, you'd have $2,000. But what if they added half a year's worth of interest after just 6 months? Then you'd have $1,500 earning 100 per cent annual interest for 6 months. So at the end of the year you'd have $2,250. That's good news, so let's keep on insisting on early additions of the interest due — and keep on racking up the cash. Quarterly interest would gain you $2,414 at the end of the year, and monthly interest would leave you with $2,613. What about daily interest? That yields $2,715. That seemed strange to Bernoulli. Calculating the interest 30 (or so) times more often only gives you an extra $102; that switch from monthly to daily is hardly worth the effort. What he discovered is that this happens because the interest gains make the total head towards a limit where there's almost no change. That limit is known as 'e'. Its value is 2.71828... There is no end to its digits, but we can define e through various mathematical expressions.

The person who did most of the groundwork on e was Leonard Euler. Euler (pronounced 'Oiler') was quite possibly the best mathematician ever to walk among us. Born in 1707 in Basel, Switzerland, Euler was almost entirely self-taught, having learned no maths at school and been rejected as a possible tutee by the prickly Johann Bernoulli, who advised him to go away and read some books. In the end, Euler became the author of the books other mathematicians would read. 'Read Euler, read Euler,' Laplace repeatedly advised his young acolytes. 'He is our master in everything.'[18]

Euler's mathematical inventions seemed to flow from the core of his being. They span a huge swath of subjects, and came so naturally to him that he continued to be immensely productive even after the loss of his sight and the gain of a gaggle of children who swarmed around his feet while he worked. Not that Euler was just a mathematician. He worked as a general advisor to King Frederick of Prussia, helping him with engineering projects, issues of artillery, and even the running of the national lottery. He also spent time working as a medical officer for the Russian Navy and carried out research in the St Petersburg Academy of Sciences. He could, it seemed, turn his hand to anything.

It was Euler who gave the base of the natural logarithm its symbol e. It is sometimes thought to be named after Euler, but few people think the man was that egotistical; it's likely that he picked it with nothing more in mind than that it was a convenient unused letter in mathematical notation. Now, though, it is often known as Euler's number, and we have calculated it to trillions of digits. What we can't seem to do is fathom the depths of this number's power. Frankly, e seems ridiculous at first glance. In 1898, for instance, the Russian economist Ladislaus Josephovich Bortkiewicz published data on 20 years of horse-kick injuries in the Prussian cavalry.[19] In 200 injury reports, it said, there were 109 in which no one died, but there was, on average, one fatality from a horse-kick every 1.64 years. Put those numbers together and you get e:

$$\left(\frac{200}{109}\right)^{1.64} = 2.71$$

If that's not raising your eyebrows, how about this: you can derive e from plotting the places the German V-1 flying bombs landed in London during the Second World War. It's also there when you trace the rate at which your DNA incurs mutations. However, none of these are accidental or mystical occurrences. They are a consequence of e's involvement in the numbers describing certain kinds of events. If an event occurs repeatedly, but rarely, and each event is independent of the others, you'll be able to describe the pattern they cause (whether it's in time or space) using something called a 'Poisson distribution'. We'll come across it again later, when we look at statistics in Chapter 7, but the way the numbers work in the Poisson distribution means that Euler's number e is always involved. However, by far the most important thing about e arises when doing calculus, because e is its own derivative. Let's have a look at why that matters.

I mentioned in the chapter on calculus that there are various rules for calculating the derivative — the slope, effectively — of a curve. When the curve has the form $y = b^x$, the rules say that its derivative is simply kb^x, where k is an unknown factor that is numerically related (in a somewhat complicated way) to b. In other words, for any exponential function, the derivative is k multiplied by the original function. Given that, there's an obvious question: is there any circumstance where the number k is equal to 1? That would be nice: it would make the function exactly its own derivative, which would make its calculus unbelievably easy.

The answer to that obvious question is yes. And that circumstance is when b, the base in our original expression $y = b^x$, is 2.71828... In other words, if you differentiate $y = e^x$, $dy/dx = e^x$.

It's almost impossible to overstate how important this makes e. If you can shift your exponential functions around, whatever base they are in, and put them in base e, you can suddenly do their calculus with consummate ease, allowing you to solve all kinds of interesting problems. If you want to know the number of new cases of a viral infection to expect tomorrow, for instance — you know that it will be somehow related to the number of cases today. It's an exponential function that you might write using arbitrary numbers. But if you want to play around with the maths and find properties such as the rate of change, it's often easier to write it in terms of Euler's number e raised to some multiple of the time that's passing. That's because that multiple will be the proportionality constant between the total number of cases and the rate of growth. For the same reasons, we use e to help us to understand a vast range of phenomena. e is central to financial matters such as compound interest; the arrangements of branching blood vessels in the human body; the way bacterial colonies grow; the rate of heat flow from a hot body to a cold one (helping shape the equations that powered the Industrial Revolution); and the natural decline in the amount of radioactive substance in a sample. To take the last example: as the original mass, m_0, emits radiation at a rate r, after time t the remaining radioactive mass is $m_0 e^{-rt}$. Much of the atomic age was built on understanding this application of e (and the relevant calculations were, of course, carried out on the slide rule).

Perhaps most importantly of all, you can work from any log table to create another table in a new base, but no base is more valuable than e. This seems to be what Oughtred was suggesting in his appendix to Napier's original book. We'll never know how he achieved such a deep insight, but he was right; as we'll see in the next chapter, e opened up the entire range of 20th-century technology — not just banking and bomb-building, but electrical innovations like the radio, the power grid, and, eventually, the computer.

Before we move on, though, we should celebrate another of Napier's achievements. In many ways, his logarithms should have been enough. As we have seen, they brought us centuries of innovation: charts of the heavens, the steam engine, the atomic age, and the slide rules that performed the mission-critical calculations of the Apollo astronauts. But Napier also brought us decimals.

The Point of Decimals

It's a testimony to the success of the decimal that we've already encountered them in this book without even having to stop and discuss what they are. Let's take a moment to do just that.

Essentially, decimals are just a different way of doing fractions. The first digit after the decimal point is the number of tenths, the second is the number of hundredths, the third is the number of thousandths, and so on. The first known use of the idea is in a 10th-century book written by an Islamic mathematician called Abu'l Hasan Ahmad ibn Ibrahim al-Uqlidisi. He even suggested a notation — effectively an apostrophe — to indicate where the decimal fractions began.

This idea of decimal fractions only came to the attention of Western mathematics scholars in 1585, when Bruges-born mathematician Simon Stevin published a pamphlet called 'La Thiende' ('The Tenth'), which explained the basics. So convinced was Stevin of the utility of decimal fractions that he said it would only be a matter of time before decimal-based coinage was commonplace.

Not with his notation, it wouldn't be. Stevin indicated the start of the decimal fractions by a zero with a ring around it. The tenths were followed by a ringed 1; the hundredths by a ringed 2, and so on. The German mathematician Bartholomeo Pitiscus cleaned up this mess in 1612 by introducing the decimal point we are familiar with today. And Pitiscus's notation was popularised by none other than John Napier in his wonderful tables of logarithms.

We now have all the ingredients in place to take our next step — an astonishing leap that takes us out of our own world, and allows us to explore others. Remember how Napier originally developed logarithms to help sailors and astronomers with their calculations? Because they were based on trigonometry (a word invented, incidentally, by Bartholomeo Pitiscus, the inventor of the decimal point) there are interesting relationships between logarithms, exponents, and the triangle-based ratios we call sines, cosines, and tangents. But this involves something we haven't met yet: the imaginary number. As we are about to see, 'imaginary' is a regrettable misnomer: this strange mathematical creature is real enough to power almost everything in the modern world.

Chapter 6

IMAGINARY NUMBERS

How we fired up the electric age

Has a mathematical invention ever been given a more misleading name? Imaginary numbers are an offshoot of algebra that formed a field — and a sphere of influence — all of their own. Though they are a different kind of number, they are inescapably real: almost nothing in the modern world works without them. The electrification of America, the innards of the mobile phone, the sound in a cinema, and the crunch of a Marshall amp all owe their essence to imaginary numbers. Silicon Valley was literally founded on them. That said, it's a good thing that the mathematician Charles Lutwidge Dodgson — better known as Lewis Carroll — didn't understand how useful imaginary numbers would be, or we'd never have encountered the Hatter's infamous tea party.

Clarence Leonidis Fender was one of millions of Americans who lost their jobs in the Great Depression of the 1930s. Fender had majored in accountancy at Fullerton College, California, and had enjoyed his work as an accountant with the California Highway Department enough to seek another job in the same field. He ended up doing the books for a firm that sold tyres. But when that job went south too, Fender decided to change things up. Indulging a childhood passion, he borrowed $600 and set up a radio repair shop.

It was 1938, and, as well as repairing radios, Fender's Radio Service offered custom-made amplifiers — public address systems, mostly — for purchase and hire. Most importantly, though, Leo Fender offered innovation. Having heard the newfangled electric guitars and their tinny amplifiers, he turned his hand to designing and building better ones. No one is quite sure how he went about it; it seems he just copied and adapted the amplifier circuits laid out in the Radio Corporation of America's *Receiving Tube Manual*, a basic 'how-to' guide for building radio equipment.

The first attempts left his workshop in 1945. The following year, he began to sell improved versions that became known as the 'Woodies' because of their hardwood cabinets. Fender's amps and guitars went on to achieve worldwide fame, and the original site of Fender's Radio Service in Fullerton is now marked by a plaque and a listing in the National Register of Historic Places. Evidently, though, Fender's early amplifiers sounded a little too like they had been designed by an accountant, because amateur electrical engineers soon began to improve them.

One of these efforts is linked to another history-marking plaque; this time, on the wall at 76 Uxbridge Road, Hanwell, in west London. The plaque says simply that this is where Jim Marshall sold his first guitar amp.

Marshall's shop mainly sold drums — he was a drum teacher — but it also sold Fender amplifiers. At the beginning of the 1960s, however, guitar players were looking for something more than the squeaky clean sound of Leo Fender's amps. Drums were getting louder, and guitarists needed amps that could rise above the noise — and maybe sound a little more interesting to boot. Marshall thought that he might make a tidy profit by designing and building his own range of amplifiers that had a distinctive, ear-splitting sound, but he didn't have the engineering chops to do it. Neither did his repairman Ken Bran. But Bran knew a kid who did.

Bran was a keen amateur radio operator, and belonged to the Greenford Radio Club, which met on Friday evenings. It was there that he had met Dudley Craven, an 18-year-old electronics apprentice at EMI Electronics in Hayes, west London. Craven was known to the club's members as an electronics genius. After a meeting one Friday, Bran persuaded him to come along to a fast-food bar for coffee, where Bran suggested Craven help out with the Marshall plan.[1]

Craven was happy at the idea of making some extra cash. And so, after work and college were over, he spent his evenings in his father's shed using his electronics nous to work out how he might improve on Leo Fender's design. He replaced some components and added new ones in search of the crunch, distortion, and monstrous volume that Jim Marshall was after. In September 1963, he knew he was on the right track when Pete Townshend, who was soon to form The Who, bought Craven's first amplifier. Townshend paid Marshall £110.

Craven's commission was less than 0.5 per cent: 10 shillings. The Marshall sound — the sound that was responsible for the birth of rock music — was born from the talents of a teenager working for pocket money. But it, and all that led to it — including the birth of radio and the electrification of America — couldn't have existed at all without imaginary numbers.

The Square Root of *What?*

Imaginary numbers are not imaginary at all. The truth is, they have had far more impact on our lives than anything truly imaginary ever could. Without imaginary numbers, and the vital role they played in putting electricity into homes, factories and internet server-farms, the modern world would not exist. But perhaps we should start this journey with an explanation of what an imaginary number is.

We know by now how to square a number (multiply it by itself), and we know that negative numbers make a positive number when squared; a minus times a minus is a plus, remember? So $(-2) \times (-2) = 4$. We also know that taking a square root is the inverse of squaring. So the possible square roots of 4 are 2 and −2. The imaginary number arises from asking what the square root of −4 would be.

Surely the question is meaningless? If you square a number, whether positive or negative, the answer is positive. So you can't do the inverse operation if you start with a negative number. That's certainly what Heron of Alexandria seemed to think. Heron was the Egyptian architect whose mathematical tricks, written in *Stereometrica*, gave us the dome of the Hagia Sophia. In the same volume, he showed how to calculate the volume of a truncated square pyramid; that is, a pyramid with the top chopped off. His solution for one example involved subtracting 288 from 225, and finding the square root of the result. The result, though, is a negative number: −63. So the answer would be found via $\sqrt{-63}$.

For some reason — whether a sense that there was some mistake, or someone copied something down wrong, or because it was so absurd — the manuscripts we have show that Heron ignored the minus sign and gave the answer as $\sqrt{63}$ instead.[2]

The square roots of negative numbers are what we now call imaginary numbers. The first person to suggest that they shouldn't be ignored was the Italian astrologer Jerome Cardano. We met Cardano during our trip through algebra, and it was here, in the solutions of cubic equations, that he stopped and stared at the issue. At first he called them 'impossible cases'. In his 1545 book on algebra, *The Great Art*, he gave the example of trying to divide 10 into two numbers that multiply together to make 40. In the process of finding those numbers, you come across $5 + \sqrt{-15}$.

Cardano didn't shy away from this unexpected encounter. In fact, he even jotted down a few thoughts about it. However, he wrote in

Latin, and translators argue about what he actually meant.[3] For some, he calls it a 'false position'. For others, it's a 'fictitious' number. Still others say he characterises the situation as 'impossible' to solve. One of his further comments on how to proceed in such a situation is translated as 'putting aside the mental tortures' and as 'the imaginary parts being lost'. Elsewhere he refers to this as 'arithmetic subtlety, the end of which ... is as refined as it is useless'. He says it 'truly is sophisticated ... one cannot carry out the other operations one can in the case of a pure negative'. By pure negative, he means a standard negative number, something like -4. He was happy with negative numbers, and wrote that '$\sqrt{9}$ is either $+3$ or -3, for a plus [times a plus] or a minus times a minus yields a plus'. And then he continued, '$\sqrt{-9}$ is neither $+3$ or -3 but is some recondite third sort of thing.' Cardano clearly thought the square roots of negative numbers were something abstruse and abstract, but at the same time he knew they were *something* — and something that a mathematician should engage with. The task wasn't for him, though; none of Cardano's subsequent writings mention the square roots of negative numbers. He left it to his fellow countryman, Rafael Bombelli, to address them a couple of decades or so later.

In what he called a 'wild thought', Bombelli suggested in 1572 that the two terms in $5 + \sqrt{-15}$ could be treated as two separate things. 'The whole matter seemed to rest on sophistry rather than truth,' he said, but he did it anyway. And we still do it today, because it works.

Bombelli's two separate things were what we now call real numbers and imaginary numbers. The combination of the two is known as a 'complex number' (it's complex as in 'military-industrial complex', speaking of combination — of real and imaginary parts — rather than complication). But let's be clear. If there's one thing we've learned in our time revisiting mathematics, it's that all numbers are imaginary. They are simply a notation that helps with the concept of 'how many'. So applying the name 'imaginary numbers' to the square roots of negative numbers is pejorative and unhelpful.

That said, we should acknowledge a distinction. What mathematicians call 'real' numbers are the numbers you're more familiar with. The 'two' in two apples; the 3.14... in pi; the fraction. And just as positive numbers are in a sense complemented by negative numbers, what we call real numbers are complemented by what we now have to call imaginary numbers. Think of them as yin and yang, or heads and tails. And certainly not as actually imaginary.

Bombelli, in his wild thought, demonstrated that this new tribe of numbers have a role to play in the real world. He set out to solve a cubic equation that Cardano had given up on: $x^3 = 15x + 4$. Cardano's solution required him to deal with an expression that contained the square root of -121, and he just didn't know where to go with it. Bombelli, on the other hand, thought he might try applying normal rules of arithmetic to the square root. So, he said, maybe $\sqrt{-121}$ is the same as $\sqrt{121} \times \sqrt{-1}$, which gives $11 \times \sqrt{-1}$.

Bombelli's great breakthrough was to see that these strange, seemingly impossible numbers obey simple arithmetic rules once they are separated out from the other, more familiar types of number during a calculation. Everything after that was just grasping the nettle.

Proceeding with Cardano's cubic equation, he eventually arrived at a solution:

$$x = (2 + \sqrt{-1}) + (2 - \sqrt{-1})$$

Separate them out into what we would now call their real and imaginary parts, and it simplifies to 2 plus 2, and $\sqrt{-1}$ minus $\sqrt{-1}$. The imaginary part disappears, leaving us with just $2 + 2$. So $x = 4$ is one of the solutions to $x^3 = 15x + 4$. Plug it in and check for yourself.

How to Imagine Reality

These days, the convention is to use i to represent $\sqrt{-1}$. The Swiss mathematician Leonard Euler (pronounced 'Oiler', remember) first came up with this. It's easy to assume that i stands for imaginary, but the truth is, as with his e, Euler may just have picked it at random. Whatever the reason, Euler's move has cemented i as the imaginary number in a very unhelpful way.

To see better what an imaginary number is, let's think of a standard number line that runs from -1 to 1 (you can think of it as a ruler placed on a table in front of you, running from -1 on the left to $+1$ on the right). We call the process of moving along the line addition and subtraction (I'm at 0.3, and I'll add 0.3 more, which takes me to 0.6). But we can also imagine making some moves by multiplication. If I start at 1, how do I get to -1? I multiply by -1. So let's picture multiplication by -1 as half a rotation, anticlockwise, around a circle (in our case, the circle passes through 1 and -1). It's actually a rotation by 180 degrees. In mathematicians' preferred units to denote angles, 180 degrees is π radians ($360°$, a whole circle, is 2π radians).

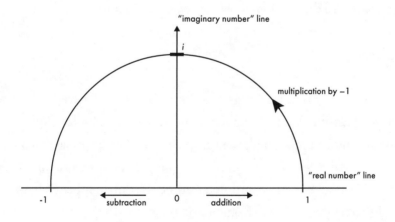

Imaginary numbers and the number line

What happens if we only do half of this rotation? It's halfway to multiplying by -1, which you can think of as the same as multiplying by $\sqrt{-1}$. That rotation, by just $\pi/2$ radians (or 90°) leaves our number up on the top part of the circle's circumference, away from the standard number line. So we can think of the square root of −1 as sitting on a number line that runs at right angles to the number line we're familiar with. It's just another set of numbers, this time on a ruler that meets your other ruler at 90° to form a cross, with +1 at the end furthest from you, and −1 right in front of you.

That leads us somewhere interesting. The link with rotation in circles means that i is related to π and the sines and cosines of angles. That relationship is mediated through the strange number e that we encountered in our last chapter. Euler worked out exactly what this looks like by taking a particular kind of infinite series (it's called a Taylor series), and deriving something now known as Euler's formula:

$$e^{\pm i\theta} = \cos\theta \pm i\sin\theta$$

This shows there is a fundamental relationship between the base of the natural logarithm and the imaginary number. What's more, you can reduce this to the relation known as the Euler identity:

$$e^{i\pi} + 1 = 0$$

To some, this is a near-mystical formula. Here we have the base of natural logarithms e; the numbers 0 and 1, which are both unique cases on the whole number line; the imaginary number, a special case all of its own; and π, which as we know is a source of power in mathematics. Despite being discovered at different times by different people looking at different pieces of mathematics, it turns out they are interrelated, coexisting in this elegant, simple equation.

Seen from a slightly different perspective, perhaps we shouldn't be surprised. As with π itself, there really isn't anything mystical about this formula. It results from the fact that numbers change and transform themselves and each other through rotations. That only happens because of what numbers are: representations of the relationships between quantities. We don't find anything mystical about moving along the familiar 'real' number line by adding and subtracting. And there's nothing different, really, about the transformations that come about through multiplications and divisions. Remember that sines and cosines are just ratios — one number divided by another — that are related to the angles within triangles, and you can represent those angles as fractions or multiples of π in units known as radians. So what we're discovering here is not some deep mystery about the universe, but a clear and useful set of relationships that are a consequence of defining numbers in various different ways.

In fact, these relationships are more than useful — they could be described as vital. Take their application to science, for example: a full mathematical description of nature seems to require imaginary numbers to exist. The 'real' numbers, of which we have learned so much, are not enough. They must be combined together with the imaginary numbers to form the 'complex' numbers that Bombelli first created. The result, says mathematician Roger Penrose, is a beautiful completeness. 'Complex numbers, as much as reals, and perhaps even more, find a unity with nature that is truly remarkable,' he says in his book *The Road to Reality*.[4] 'It is as though Nature herself is as impressed by the scope and consistency of the complex-number system as we are ourselves, and has entrusted to these numbers the precise operations of her world at its minutest scales.' In other words, imaginary numbers had to be discovered because they are an essential part of the description of nature.

The greatest scientific impact of imaginary numbers is perhaps seen in their central role in the Schrödinger equation, which governs the operations of quantum theory. This mathematical endeavour is our best means of describing and predicting the behaviours of some of the basic constituents of nature. Whether it's photons (packets of electromagnetic energy such as light), electrons (the negatively charged parts of atoms), protons and neutrons (which constitute the nucleus of atoms), or the various forces that act between such particles, everything in nature seems to abide by the laws wrapped up in this equation. It was formulated by the Austrian physicist Erwin Schrödinger in 1925, and won him a Nobel Prize in 1933. It remains one of the most important, and most concise, statements about the behaviour of the natural world, and it's worth taking a glance at it, even though we don't need to unpack it. See how i, the imaginary number, is front and centre:

$$H\Psi(r,t) = i\hbar \frac{\partial}{\partial t} \Psi(r,t)$$

Much has been said about the strangeness and impenetrability of quantum theory. The American physicist Richard Feynman, who won a Nobel Prize for his theory of quantum electrodynamics, once declared, 'I think that I can safely say that nobody understands quantum mechanics.'[5] It certainly is strange: the Schrödinger equation unleashes phenomena such as 'superposition', where subatomic particles seem to have a kind of multiple existence where they can, effectively, be in two places at once, or move in two different directions at once. Then there is 'entanglement', a phenomenon that Einstein referred to as 'spooky action'. It is so spooky because entangled quantum particles seem to have an effect on each other's properties, no matter how widely separated they are in space. Give one a kick, and its entangled twin will be found to have been affected by that kick, even if it is half a universe away.

All of the strange quantum properties were spotted in the equations of quantum theory long before they were observed in experiments. And those equations involve complex numbers.

This is because we are actually dealing with phenomena that are best described as waves. The first person to do this for quantum stuff was the French aristocrat Prince Louis de Broglie (pronounced 'Broy'). He was a PhD student at Paris University's Faculty of Sciences when he came up with the idea that everything we think of as a particle can also be represented as a wave, and vice versa.

De Broglie's principal idea was to treat an electron inside an atom as a wave whose wavelength depends on its energy. When it is given more energy, its wavelength decreases. The only way to write a full mathematical description of this wave is to use complex numbers. The most important factor in that description is something called its phase. Phase usually gives us a measure relative to something: the phases of the Moon, for instance, are the result of the relative position of the Sun, Moon, and Earth. In physical waves, such as water waves or sound waves, the phase is a measure of where you are relative to the start and end of the wave's cycle: halfway through, say. In quantum waves, though, the phase is an entirely different beast: it's a straightforward property of the quantum particle. You might describe the electron as having a particular position, a particular momentum, and a particular phase. Strangely, though, this quantum phase doesn't actually exist in the same physical space as the particle itself.

To make this idea work alongside Einstein's relativity, de Broglie had to introduce the idea of extra dimensions. If the wave carried a particle's energy and momentum through physical space, it would involve movement faster than the speed of light, and relativity forbids this. So de Broglie set things up so that it is a 'phase wave' rather than a 'matter wave'. Here, believe it or not, the wave is an undulating complex number that oscillates in an abstract dimension.

That might sound mad to you already, but it gets worse. Quantum physics ascribes an extra dimension to each physical property of each electron. That's why Erwin Schrödinger described his extension of de Broglie's innovation as 'the multi-dimensional wave mechanics'.[6] That little i in the Schrödinger equation looks so simple, but it creates a huge, complex (in every sense of the word) landscape composed of an almost infinite number of dimensions.

This multidimensional landscape is used to construct something known as 'Hilbert space'. It is named after the mathematician David

Hilbert, whose study of calculus and geometry led him to introduce the idea of an arena where the three dimensions of space that we are familiar with have branched out into an infinite number of dimensions. This is the root of the 'many worlds' interpretation of quantum theory, which claims that there are, effectively, alternative universes to our own, each of which contains a subtly different version of our reality.

But, astonishingly, the quantum many worlds idea is not even the most complicated outworking of the complex numbers. That's because quantum mechanics is not the final theory of the universe. For that, we might need a set of complex numbers known as the quaternions and (maybe) their cousins, the octonions. It's time to travel to Dublin, to meet the bridge-defacing mathematician Sir William Rowan-Hamilton.

Putting the *i* in Alice

Let's start with the Pythagoreans, whom we met several chapters ago. These were the scholars who entered their school of study through an archway that declared 'All Is Number', and may or may not have drowned one of their colleagues for revealing a flaw in their belief system.

This fanaticism started with a love and respect for music. Greek music — the music of the cosmos, in Pythagorean minds — could be reduced to ratios of numbers such as 1 to 2, 3 to 2, and 4 to 3. Two strings of equal tension whose lengths are in the ratio 1 : 2 produce notes an octave apart. Those with the ratio 3 : 2 are a 'perfect fifth' apart. If the length ratio is 4 : 3, this is the interval known as a fourth. The numbers 1, 2, 3, and 4 were sacred to the Greeks, and their sum was 10, the perfect number. They represented it as a triangle of symbols. This was the set of four, the 'tetractys'. They used it when they swore oaths: 'by him that gave to us the tetractys, which contains the fount and root of ever-flowing nature'. In other words, they were deadly serious about numbers, believing them to be the key to understanding the entire universe.

The Greek tetractys

And they may not have been wrong. In 1960, a Hungarian mathematician called Eugene Wigner wrote an essay entitled 'The Unreasonable Effectiveness of Mathematics in the Natural Sciences'.[7] His point was simple: it seemed to Wigner that our invention of the concept of numbers, together with a set of rules for their manipulation, has allowed us to describe and predict any number of real-world phenomena. But, he said, numbers and their rules — mathematics — are a product of the human brain; why should they give us such insight? As Wigner put it, 'the enormous usefulness of mathematics in the natural sciences is something bordering on the mysterious'. There is, he adds, 'no rational explanation for it'. He calls it a 'miracle'.

Sixty years later, the miracle is ongoing, but it has expanded out of the realm of the natural numbers that so intrigued the Pythagoreans. As we have seen, it has extended into the realm of the complex numbers that allowed us to create amplifiers, and pin down the behaviours of subatomic particles. But it didn't even stop there, because now we are coming into sight of what are known as the 'hypercomplex' numbers. It sounds daunting, I know, but bear with me. It is worth bravely venturing into this landscape, because the rewards are an extraordinary post-Pythagorean insight into what the universe might really be made of.

The hero of this adventure is an Irish mathematician. William Rowan Hamilton was born in Dublin in 1805. His biographers might have been overdoing it when they said he was so clever that he mastered ten ancient languages — including Chaldee, Syriac, and Sanskrit — by the time he was 10 years old. However, it is true that he was a boy genius: when he read Laplace's newly published treatise on celestial mechanics at 17, he found a mistake that no one else had noticed. By 22, he was Ireland's astronomer royal. By 30, he had been knighted for his services to scientific advancement.

That was in 1835, the same year Hamilton became obsessed with complex numbers. His heart's desire was to take them further. If, he reasoned, the imaginary number i gave us another dimension of number space, who was to say there weren't more dimensions waiting to be discovered? He decided to experiment with inventing two additional sets of numbers — two more dimensions of the number line, effectively. He denoted them j and k, and he began to test whether he could do arithmetic with them in the same way that Bombelli had done it with the square root of -1 (now known as i) three centuries before.

As it turned out, Hamilton could ascribe mathematical properties to j and k that would allow addition and subtraction between the 'triplet', as he called it, of i, j, and k. But he couldn't manage multiplication or division. His dedication to expanding the set of acceptable operations is reflected in his children's question to him every morning. 'Well, Papa,' they would ask their father when he came down to breakfast, 'can you multiply triplets?' Hamilton would reply with what he termed 'a sad shake of the head'.[8]

And then he cracked it. Hamilton's moment of insight is a famous anecdote in the history of mathematics. On 16 October 1843, he was walking along Dublin's Royal Canal with his wife when he suddenly realised what relation between i, j, and k would resolve the problem. 'I then and there felt the galvanic circuit of thought close; and the sparks which fell from it were the fundamental equations between i, j, k,' he later said. He was so excited, and so determined not to let the thought slip away, that he carved the relation into the stone of a nearby bridge.

Hamilton's hasty act of vandalism has long since disappeared, eroded by time and touch. Today, a plaque marks the spot of Hamilton's enlightenment, and its inscription includes the spark-inducing relation:

$$i^2 = j^2 = k^2 = ijk = -1$$

Hamilton added his triplet to the familiar 'real' number set to create what he called the quaternions. He said he derived this English-language name from the concept of the Greek tetractys, the original, mystical set of four. It was not an accidental connection. Hamilton was convinced that scientists were cut from the same cloth as the ancient Greek thinkers. They should, he said, 'learn the language and interpret the oracles of the universe'.[9] He even shared the Greeks' admiration for poetry, counting the Romantic poets Wordsworth and Coleridge among his closest friends. In Hamilton's view, there was a desperate need to reconnect science to philosophy and the search for the divine. And with the quaternions he felt he had taken the first step.

The day after making his discovery, Hamilton wrote to his friend John Graves, a lawyer who had shown more than a passing interest in algebra. Graves's response was audacious: why stop there? 'I have not yet any clear views as to the extent to which we are at liberty arbitrarily to create imaginaries, and to endow them with supernatural properties,' he wrote in a reply on 26 October. Two months later, Graves wrote again. He had doubled the imaginaries again, and created the mathematics of the octonions. As if to satisfy the musical tastes of the ancients, these two men had created an octave of numbers.

The pair tried to go further, but failed. That, we now know, is because it is impossible. Nature, it seems, is written in sets of numbers, but not infinitely many of them. Mathematicians have proved that, with the octonions, we now have the full set of possible systems with which humans can perform the unreasonably effective work of describing the universe in numbers.

So how does it work? You won't be surprised to learn that it's complicated. So complicated, in fact, that it seems to have inspired one of the greatest scenes of absurdist fiction in the English language: the Hatter's tea party in Lewis Carroll's *Alice in Wonderland*.

Lewis Carroll is the pen name of the Oxford University mathematician Charles Lutwidge Dodgson. Dodgson was a tutor in geometry at Christ Church College. Rather conservative by nature, his favourite book was Euclid's *Elements*, and in a book called *Curiosa Mathematica* he eulogises his hero's pure mathematics: 'The charm lies chiefly, I think, in the absolute certainty of its results: for that is what, beyond almost all mental treasures, the human intellect craves for. Let us only be sure of something!' Dodgson, it's safe to say, was not really a fan of experimental ideas — or progress.

His most famous book began its life as the rather dull 1864 manuscript *Alice's Adventures under Ground*. It did not contain any kind of tea party. However, around the time of writing, Dodgson was becoming more and more frustrated by the directions his beloved subject was taking. Euclid's algebra was becoming old hat; the new hat had the shape of abstract algebra and, in particular, complex numbers and the quaternions. Dodgson wrote of his concerns to his sister, talked them over with colleagues and conveyed them in articles for mathematical journals, but no one seemed to be listening. And so he employed one of Euclid's favourite rhetorical tricks: the reduction to absurdity.

Alice in Wonderland is peppered with Dodgson's pokes and jabs about his least favourite mathematical trends.[10] There are subtle digs at negative numbers, symbolic algebra, and a field called 'projective geometry' and its 'principle of continuity' (to parody its ideas, Dodgson has a baby turn into a pig). The English literature expert Melanie Bayley, who gathered the pieces of the puzzle and put them together, thinks Dodgson would have particularly enjoyed managing to smuggle these references into the house of Henry Lidell, the dean of Christ Church college.[11] Liddell was the father of Alice, the girl who was the protagonist in Dodgson's original story. Bayley has uncovered documents showing that Dodgson was angry about the introduction of symbolic mathematics into the Oxford syllabus, and had a private row with Dean Liddell on the subject, right around the time he was writing *Alice*. Bayley imagines Dodgson slipping his arguments into the revised copy of the book that he gave to the Liddell family, so that his objections would sit on the Dean's drawing room table as a secret joke — or perhaps one that was shared only with Dodgson's sympathisers.

For all his problems with symbolic algebra, it was Hamilton's quaternions that inspired Dodgson's most scathing attack. At the Hatter's tea party, Alice comes across three strange characters: the Hatter, the March Hare and the Dormouse. Alice notes that they 'keep moving round'. This seems to be a reference to one of Hamilton's greatest innovations: his means of finding a way for the quaternions to multiply and divide. It can be summed up in the diagram below.

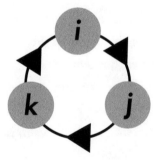

William Hamilton's cycle of quaternion multiplication

What this says is that, when it comes to quaternions, the order in which you multiply things matters. We say 2×3 is the same as 3×2. But $i \times j$ (which equals k) is not the same as $j \times i$ (which gives $-k$). It matters *how* you move around the circle. Alice's conversation with the March Hare is a consequence of Dodgson's mistrust of this new mathematics: 'say what you mean' is, as the Mad Hatter points out, not the same as 'mean what you say'.

Meanwhile, a fourth character, Time, is missing from the party. His absence has resulted in the problem that it's always six o'clock — always teatime. Here, Dodgson seems to be reacting to Hamilton's great contention that the quaternions are intimately linked to the physicists' problem of representing time. In 1835, Hamilton had written a book called *Algebra as the Science of Pure Time*. His invention of the quaternions resulted in him suggesting that one of the four numbers is time. In some of his own (somewhat eye-watering) poetry, Hamilton noted how 'the One of Time, of Space the Three,/Might in the chain of symbols girdled be'. It was an anticipation of time as the fourth dimension — not the kind of time that passes, but the kind that exists, static and absolute: 'the before and after; precedence, subsequence, and simultaneity; continuous indefinite progression from past through the present to the future', as he put it. Hamilton was quite the philosopher, it turns out. 'There is something mysterious and transcendent involved in the idea of Time,' he wrote, 'but there is also something definite and clear: and while Metaphysicians meditate on the one, Mathematicians reason from the other.' It's a bit wordier than some of the quotes about time from its other great exponent, Albert Einstein. One in particular comes to mind: Einstein once said that 'The only reason for time is so that everything doesn't happen at once.' But it's essentially the same point: time, in the mind, is just an illusion. And, as he shows in his scene, Dodgson was having none of it: in the absence of Time, there is no progress.

With relativity, Einstein, not Hamilton, was the one to prove it. And Einstein didn't even use Hamilton's four-dimensioned quaternions to develop his special and general theories of relativity, which describe the properties of space and time, and how things are allowed to behave as they move through them. There had been a war in mathematics, something akin to the video format wars between Betamax and VHS. In the end, the quaternions had lost out to a mathematical innovation known as 'vectors', which specify numbers through a direction and a distance on the numerical equivalent of a navigational map. Ever after, quaternions were the poor relations of the vectors. But, despite Einstein's use of four-dimensional vectors, we can still laud Hamilton for putting the idea of a fourth dimension firmly into the hearts and minds of anyone looking to unlock the cosmos of the Pythagoreans. And while the quaternions have seen very little real-world use, their extension, the octonions, is a strong candidate for unlocking a final theory of physics.

The Eight-Fold Way

Even William Rowan Hamilton, tireless champion of his quaternions, baulked at promoting the octonions. Four-dimensional algebra made it possible to account for time. But to what possible use could you put eight-dimensional algebra? What was anyone going to do with all that extra space? Especially when its mathematical rules were so convoluted.

The quaternions had a rather simple mathematical relationship. But when Graves worked out that you could perform arithmetic in eight dimensions, he had to break new — and seemingly ridiculous — ground. Never before, for instance, had the position of parentheses mattered. But with the octonions $3 \times (4 \times 5)$ was not the same as $(3 \times 4) \times 5$.

When dealing with standard numbers, mathematicians perform the calculations within parentheses first. They would see $3 \times (4 \times 5)$ as 3×20, for instance: 60. But it would make no difference to them if the parentheses were moved. They would calculate $(3 \times 4) \times 5$ as 12×5 and still get 60.

With the octonions, however, normal rules no longer apply. While quaternions use three numbers i, j, and k, octonions use the seven numbers $e_1, e_2, e_3, e_4, e_5, e_6$, and e_7. In case you were wondering, e_1, e_2, and e_4 are comparable to the quaternions i, j, and k. The diagram overleaf shows how they work together for mathematical purposes.

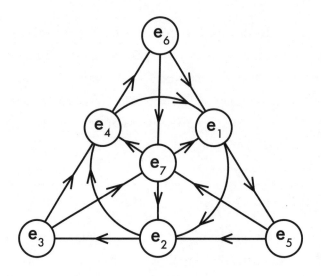

The Fano plane for octonion multiplication

It's beautiful in a way. This map of the octonion landscape is called the Fano plane, after the Italian mathematician Gino Fano. It has a mystical quality, reminiscent of the all-seeing eye — the Eye of Providence — on the reverse of the Great Seal of the United States and on the dollar bill. Mathematicians of the Renaissance would look at this and see the eye of God contained within the triangle that represents the Christian holy trinity. Their modern equivalents — or some of them, anyway — see something different: a shorthand description of how the universe holds together.

This is not a piece of mathematics that has yet changed human history. The application of the octonions is very much a work in progress that may or may not lead somewhere. But the teasing power of these strange numbers, the way their properties reflect what we know of how the forces and particles of nature work, is enough to send some physicists sliding down the rabbit hole.

We have already mentioned some of the odd properties of quantum theory. When we describe how the various subatomic particles operate, some of the strangeness in the maths reflects the properties of the quaternions. Take the Heisenberg uncertainty principle. This says that certain pairs of a particle's properties — position and momentum, for instance — can't both be precisely known at the same time. This is a result of the fact that, in quantum mathematics, the order of things matters in the same way that it does with the quaternions i, j, and k.

The unresolvable strangeness of quantum theory led both Einstein and Schrödinger to walk away from it. They had worked together to expose its flaws, and tried to convince others — particularly Niels Bohr, often considered the founding father of quantum theory — that it would be better to start again. In 1939, Einstein gave a lecture to an audience that included Bohr. As he looked him dead in the eye, Einstein said his goal was now to replace quantum mechanics.[12]

Schrödinger also walked away from developing quantum theory any further, at almost the same time as Einstein. Both set to work — independently — on a theory that would unite quantum physics and relativity. The idea was to create a grand, final theory that encompassed both relativity's take on the cosmic properties of the universe and quantum theory's description of the subatomic world and its forces. This would give a single mathematical description to the entire cosmos. Neither of them succeeded, and they managed to fall out spectacularly over their public excoriations of each other's work.[13] After Schrödinger gave one particularly clumsy press conference about the poor state of Einstein's thinking, Einstein snubbed his former friend, refusing to reply to any of his letters for three painful years (for Schrödinger, anyway).

Others have taken up the baton now, but no one would claim to be close to achieving the goal; mathematicians and physicists — and those who work the fertile fields in between the two — are still exploring a number of different paths. What's rather lovely, for our purposes, is that the mathematics of imaginary numbers, the octonions in particular, are now a prime source of optimism.

It started with string theory. This is an attempt to build all the particles and forces of physics by starting with nothing more complicated than vibrating strings of energy. The strings vibrate in one way, and we have an electron. A different vibration gives us the electromagnetic force. It's highly redolent of the idea that the mathematics of music is entwined with the mathematics of the cosmos; the Pythagoreans would love it.

However, this approach only works if we invoke 'extra' dimensions of space (these are a different set of extra dimensions from those introduced by Louis de Broglie). According to string theory, there are perhaps seven hidden dimensions to be added to the three we inhabit. And in this scheme, the properties of matter relate together in ways that can be mathematically defined using the octonions. While string theory is unlikely to be the final answer, it is probably our best effort at a 'quantum theory of gravity' so far, and it is hinting that the final theory, whatever it turns out to be, could well involve octonion mathematics.

Those hints come from the way that particle physicists assemble their menagerie of particles. The 'Standard Model' is a kind of zoological classification system that puts each of the particles into a group with others that have similar properties. One group is the hadrons, for instance, which are all composed of the quarks that we met when we looked at algebra. Hadrons have electrical charge that is a multiple of the charge on the electron (the multiple can be zero). You might be familiar with the protons and neutrons that sit in the nucleus of an atom: these are hadrons. There are many other groups, including the leptons (where electrons reside) and the bosons (such as the Higgs boson).

The various different classifications, properties, and behaviours of these particles make the Standard Model something of a mess. We struggle to make sense of where all its rules come from. But there are signs that the mess only looks like a mess because we haven't yet found how it all maps to the intricacies of the Fano plane, and where the gravitational force comes in. As the Abel-prize–winning mathematician Michael Atiyah has put it, 'The real theory which we would like to get to should include gravity with all these theories in such a way that gravity is seen to be a consequence of the octonions ... It will be hard because we know the octonions are hard, but when you've found it, it should be a beautiful theory, and it should be unique.'[14]

As yet, of course, this is a hypothetical application of imaginary numbers. But there is another beautiful theory that is immensely practical and has been in use for more than a century. It was brought to us by a German-born man called Charles Proteus Steinmetz. And it is this story that tells us, perhaps more than any other, just what a debt our civilisation owes to mathematics.

The Electrification of America

There'll be some familiar eccentrics in this tale. There's the cranky, socially awkward Thomas Edison, sometimes known as the inventor of the electric light bulb and the Wizard of Menlo Park, the New Jersey district where he opened his first laboratory. There's Nikolai Tesla, often portrayed as a mad genius obsessed with creating breathtaking displays of lightning with byzantine electrical installations. There's Michael Faraday, the deeply religious son of a blacksmith and creator of the first electric motor. And the Scottish pioneer of electromagnetic theory, James Clerk Maxwell — known as 'Dafty' to his school friends because of his weird-looking home-made shoes. However, our central character was more eccentric than any of them.

Karl August Rudolf Steinmetz was born in Breslau, Prussia, in 1865. He suffered from the same condition as his father and grandfather: kyphosis, a disorder of the spine that causes a rounded, hunched, back. He grew to be somewhere around 4 feet 9 inches tall, but the stoop made him seem much smaller. Steinmetz's intelligence was extraordinary, and he flew through his education. Fellow students recognised his gift, and paid him handsomely for private tutorials. They gave Steinmetz the nickname Proteus, after the shape-shifting Greek god who could impart wisdom to those who touched him. When he became involved with an outlawed socialist group who dreamed of an end to poverty, equality for all, and freedom from the ruling classes, Steinmetz was hunted by the authorities and fled to America. Here, the 24-year-old Karl resurfaced as Charles Proteus Steinmetz. Steinmetz arrived in America in 1889. By the end of 1893, his genius had changed the American way of life.

In 1821, Michael Faraday had come up with the idea behind the electric motor. His original apparatus involved a bowl of mercury, a magnet, a battery and a stiff wire. The interplay of electricity running through the wire with the magnetic field causes the wire to move in a circular path around the magnet. Within a few months, engineers had taken Faraday's invention and created what we would recognise as electric motors. Within a decade, inventors had turned the process on its head to generate electricity in a wire rotating around a magnet. By 1882 we had the electric telegraph, the telephone, electric lighthouses, and electrical power stations. What we didn't have was a reliable, efficient way to get the power of electricity out to homes and factories.

The main problem was simply that too much of the electrical energy would be lost when transporting it from the power station. Less power is lost when using the 'alternating current' (AC) that cycles in smooth sine curves between positive and negative values of the current, but to Thomas Edison, that was another problem. Edison had made significant investments in direct current (DC), the constant electrical power you get from a standard battery. He had essentially bet the family farm on a whole slew of DC circuits, switches, and light bulbs, and so championed DC power as the best option, suggesting that small DC generators could be installed in every building to minimise the energy loss when it is transported. Unfortunately for Edison, the board of his company, Edison General Electric, felt that this would be a mistake.

Their main competitor, Westinghouse Electric, was a well-resourced infrastructure company looking to create out-of-town power plants, such as a hydroelectric generator at Niagara Falls. Westinghouse favoured AC, partly because the barnstorming genius Nikola Tesla had already designed a complete AC power network that could serve a city — maybe even a nation. The ever-stubborn Edison fought hard for DC, and got himself thrown off the board; Edison General Electric became General Electric. And then GE went all out for AC. They began to buy up companies with the experience, engineers, and patents they needed. One of those companies was the employer of Charles Proteus Steinmetz.

Steinmetz was already known to be a big player; he had done industry-changing work on stemming the energy losses that arise when electrical power is converted between different voltages. But in the year that he became a GE employee, he wowed even those who were already fans. How? By embracing imaginary numbers.

Steinmetz's big moment came in August 1893, at the International Electrical Congress. It was part of the 1893 World's Fair in Chicago, the entirety of which was powered by Nikola Tesla's AC generator. Steinmetz had been invited to address the Congress, and had thought hard about what electrical engineers needed in order to make progress on rolling out electricity for all. In the end, he advocated not new hardware, but new tools for thinking.

Word of what Steinmetz was proposing had already leaked. Before he began his talk, the session chair, a Professor Henry Augustus Rowland, prepared the audience. 'We are coming more and more to use these complex quantities instead of using sines and cosines, and we find great advantage in their use,' he said. 'Anything that is done in this line is of great advantage to science.'[15]

Complex quantities, these days, are known as complex numbers: the combination of a 'real' and an 'imaginary' number, such as $5 + \sqrt{-15}$. In his talk, Steinmetz proposed that electrical engineers use complex numbers in all their calculations and designs. The idea was an immediate hit.

Without complex numbers, engineers working on AC were hitting a brick wall. As a generator turbine rotates, driven by something like the tumbling water at Niagara Falls, the electricity it produces has a kind of rotation too. Think of the electricity as a point on the rim of the generator wheel: as the wheel turns, the way the height of that point above or below the axle (the zero point) varies traces out the smooth up and down curve of a sine wave. AC changes in the same way, its current smoothly alternating between positive and negative. That is, it rises from zero to a maximum value, then decreases back to zero and switches its polarity until it reaches the negative maximum (the minimum) and moves back towards zero.

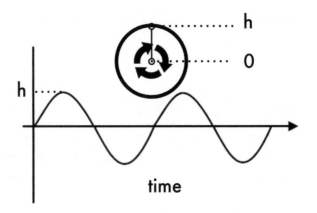

The height of a point on a rotating wheel traces out a sine wave

This cycle continues for as long as AC flows, which means that if you want to know the current's value, you have to know where you are in the cycle of the sine wave. In other words, time is a factor. In itself, that isn't a problem: we can describe an AC generator's changing currents (and voltage) using the language of sines and cosines. But once we introduce circuit elements such as the switches, resistors, capacitors, and inductors that are used in household electrical circuits or in manufacturing machinery, the mathematics involved becomes a heinous mess because they change the 'phase' of the various waves. Phase gives us a measure of how far through its cycle a wave has progressed, but when you introduce a phase shift with a capacitor or an inductor, the algebra becomes deeply painful to deal with. That's because the various waves are now at different points in their cycle at any given time. As a result, engineers were overwhelmed by the calculations they needed to perform right from the start of the electrical age: all of the various values of interest depended on the time at which you analysed your circuit. But Steinmetz removed time from the equations — just by switching from sines and cosines to complex numbers. In some ways, it was quite a simple innovation. He showed mathematically that finding the sum of sinusoidal curves could be seen as adding two complex numbers together using (like Bombelli) the appropriate arithmetic. It involved Euler's work relating sines, cosines, and imaginary numbers:

$$e^{ix} = \sin(x) + i \cos(x)$$

Using this relation, engineers no longer needed to worry about how all the phases varied while their AC circuits operated. Complex numbers gave them an easily acquired snapshot of the circuit's performance at any moment; the time-varying parts of the circuit calculations disappeared. This was how Steinmetz put it in the opening to his talk: 'Where before we had to deal with periodic functions of an independent variable, time, we have now to add, subtract, etc., constant quantities — a matter of elementary algebra.'

It changed everything. When engineers calculated the influence of capacitors (which store energy by creating an electric field) and inductors (which temporarily store energy in an electrically induced magnetic field) on an electrical circuit, they could now simply include their contribution as 'imaginary' parts to the equation. Analysis using complex numbers also resolved puzzling phenomena such as why certain circuits seemed to have surges in the total alternating current when the circuit branches and splits the original current flow in two. The complex numbers made it clear that the surges were due to the engineers' current-measuring equipment responding to AC.

Suddenly, electrical engineering was easy — or a lot easier, anyway.

Within a few years, the industry had taken over the world, and Charles Proteus Steinmetz was a hero and a celebrity. He would hang out with Tesla, Einstein, Edison, Marconi, and countless other famous names. Like them, he was considered an eccentric. He kept alligators, black widow spiders and a Gila monster at his home. He could often be seen in a canoe, surrounded by books and papers, while floating down the Mohawk River. His love of playful shenanigans earned him the moniker 'the Wizard of Schenectady' after he built a 120,000-volt lightning generator at General Electric's research labs and used it to destroy a specially constructed model village (*The New York Times* headlined their story about this escapade 'Modern Jove Hurls Lightning at Will').[16] But for all the colour of his character, his reputation was earned and maintained through his genius. There is a lovely story, published in a 1965 issue of *Life* magazine, that suggests just how highly respected he was.[17] Henry Ford's fledgling car manufacturing company was once having trouble with one of the generators that powered the production line. They called Steinmetz in to consult on the problem, and he solved it by lying down in the room where the generator was housed. For two days and nights he listened to its operation, scribbling calculations on a notepad. Eventually, he got up, climbed up onto the giant machine, and marked a point on the side with a chalk cross. He descended, and told the engineers to replace 16 of the generator's wire coils — the ones behind his chalk mark. They did what they were told, turned the generator back on, and discovered to their utter astonishment that it now worked perfectly.

That story alone would be enough, but it gets better. From their headquarters at Schenectady, New York, General Electric sent Ford a $10,000 invoice for Steinmetz's services. Ford queried the astronomical sum, asking for a breakdown of the cost. Steinmetz replied personally. His itemised bill said:

Making chalk mark on generator $1.
Knowing where to make mark $9,999.

Apparently, the bill was paid without further delay.

Once the transition to complex numbers was adopted, the electrical industry never looked back. Now its engineers could design grids and components, generators and transformers, knowing exactly how they would behave. It was, as Henry Rowland said, of great advantage to science: the widespread availability of electric power changed what could be done in a laboratory, and humanity saw a huge upswing in the volume of work coming out of scientific institutions. Within a few years we saw the invention of radio and television transmission and of cathode ray tubes. Radio engineering led to the amplifier circuits that Leo Fender and Jim Marshall would later turn into cultural and commercial landmarks. And this was the era not only of General Electric and Westinghouse, but also of Bell Labs, AT&T, the Philips and Osram light bulb companies, and the International Business Machines Corporation (IBM). The year 1901 saw the patent of the first semiconductor device, for example, and within a few years diodes and triodes had arrived — these are some of the most significant components of the electrical and electronic machinery that would pour out of the new hi-tech companies over the next few decades. Let's close this chapter by examining one innovation in particular: a circuit that creates and amplifies audio-frequency signals. It doesn't sound like much, perhaps, but it brought us Silicon Valley.

Imaginary Numbers Make Real Money

If you visit 367 Addison Avenue in Palo Alto, California, you'll find yet another plaque. This one is in front of a garage that is on the National Register of Historic Places because, as the inscription says, it is 'the birthplace of the world's first high-technology region'.

The garage belonged to David Packard; it was where he and his friend William R. Hewlett set up their electronics company, based on Hewlett's design for an 'audio oscillator'. Hewlett had developed the oscillator — a sound generator, effectively — as part of his master's degree in electrical engineering at Stanford University. He submitted his thesis, 'A New Type Resistance-Capacity Oscillator', on 9 June 1939. It is just 15 pages long, and describes the design as light and portable, simple to manufacture and use, combining 'quality of performance with cheapness of cost to give an ideal laboratory oscillator'.

The appendix of Hewlett's thesis makes interesting reading for us. Bearing in mind that the notation of electrical engineers uses 'j' instead of 'i' (because electrical current is already designated as i), you can see the importance of complex numbers. Hewlett lays out the central aspects of his oscillator's performance using a few lines of equations that host a blizzard of js.

Hewlett's advisor on the project was Frederick Terman, who was encouraging all his students to set up their own companies on the west coast rather than travel to the east coast where all the electrical and electronic action — with Bell Labs at the centre of it — seemed to be taking place. With Packard's help, Hewlett did just that, establishing his garage as the first fabrication facility of what came to be known as Silicon Valley, and establishing complex numbers as the central pillar of west coast technology. Hewlett and Packard marketed their oscillator as the 'HP200A' so that it didn't sound as if they had only just developed their first product. That morphed into the HP200B. The Walt Disney company bought eight of these to use in a new, exciting project: an innovative animated movie called *Fantasia*. This was how Hewlett and Packard found themselves at the heart of an entertainment revolution.

Walt Disney's *Fantasia* was launched with 'Fantasound', a technology designed to faithfully re-create the sound of a symphony orchestra playing in the cinema. It involved using an array of intricate electronics systems — including the Hewlett–Packard devices — and when the movie was released in the 1940s, it showed just what could be done with sound using 'electronic amplifiers' such as the one buried in the HP200B. Hewlett and Packard suddenly had a reputation.

They didn't yet have much money, though: their first year's trading made the pair just $1,563 in profit — around $30,000 in today's money.[18] The business grew during the years of the Second World War — and the US military awarded the company an 'E Award' for excellence in its imaginary-number-inspired products. By 1951, sales had reached $5.5 million — today that would be around $57 million. When Packard died in 1996, he left a personal fortune of over $4 billion. Most of Packard's money went to charity; Hewlett and Packard were always generous with their profits, their time, and their resources. One beneficiary of their largesse was a 12-year-old boy called Steve Jobs, who enjoyed an internship with HP in the summer of 1967. That was his first step towards founding Apple Computers with Steve Wozniak — who had been working for HP as a calculator designer. You can trace the origins of corporate giant Apple — one of the richest companies in the world — directly to the power of imaginary numbers.

In the 21st century, it's now impossible to overstate the impact of imaginary numbers on your daily life. Radio broadcasts, guitar amplifiers, and cinema surround-sound systems are only the start of their cultural legacy. Most of our essential digital tools are reliant on some kind of complex number processing. Take a look at your mobile phone, for example. Its MP3 music files are created using a mathematical technique called fast Fourier transforms, which involves calculating with complex numbers (we'll get to Fourier transforms in the next chapter). The same technique controls the way the signal is beamed from the base station to your phone. Designing a battery for your phone involves modelling the way it generates heat using complex numbers. Lighting up the screen's display — determining which pixels show which colours, and at what intensity — relies on complex number routines too. There's more than one 'i' in iPhone, it turns out.

If you were one of the students who complained to your maths teacher that there was no point in anyone learning how to use imaginary numbers, maybe it's time to think about putting down your phone, turning off your music, pulling the wires out of your broadband router, cutting off your domestic electricity supply, and never going to a cinema or a concert hall again. Or maybe you could just admit you were wrong.

It may be that you didn't learn about imaginary numbers, of course; school curricula change over the decades, and topics come and go. One that is definitely coming — and hopefully will stay — is statistics. The subject is often maligned: 'There are three kinds of lies: lies, damned lies, and statistics', as a popular aphorism has it. Mark Twain once declared that 'Facts are stubborn things, but statistics are pliable'. Even the great physicist Ernest Rutherford is supposed to have cast aspersions on them: 'If your experiment needs statistics', he is thought to have said, 'you ought to have done a better experiment'. It's all very unfair, as we're about to see.

Chapter 7

STATISTICS

How we made everything better

*In the public imagination, it's a byword for chicanery, but statistics
is about the search for truth. It may not always be pretty, but it is
an invaluable Swiss Army knife of mathematical tools — probes,
tweezers, scalpels, and scrapers — that can expose the true meaning
of data. It established a few famous names — Guinness, Florence
Nightingale, JPEG — but buried countless others whose work didn't
stand up to its scrutiny. Without statistics, we'd be buying quack
medicines, ignorant of the benefits of vaccination, and unable to
stream movies and music. In other words, we'd be living shorter and
less enjoyable lives. Whether the benefits are enough to compensate
for its dark and disturbing origins and the wicked works of its early
adopters is a decision that only you can make.*

In 1662, the last year anyone saw a dodo, a London draper published
the first statistical analysis of a citizen's chances of imminent death. It
was the time of bubonic plague, and King Charles II was keen to create
some kind of warning system that would make Londoners aware of

the growing risks. John Graunt decided he might be able to help. He found his raw data in the published but largely unread *Bills of Mortality*, a weekly digest of deaths.[1] 'Finding some Truths, and not commonly-believed Opinions, to arise from my Meditations upon these neglected Papers, I proceeded further, to consider what benefit the knowledge of the same would bring to the World,' he wrote in the preface of *Natural and Political Observations Made upon the Bills of Mortality*. He was determined, he said, not to engage 'in idle, and useless Speculations: but … present the World with some real Fruit from those airy Blossoms'. No one had ever before referred to the death lists of the *Bills of Mortality* as 'airy Blossoms'. But then neither had anyone actually analysed the population's chances of surviving the coming week.

Graunt was a likeable man, by all accounts: hospitable, studious, intelligent, and generous. He once gave the diarist Samuel Pepys a viewing of his architectural prints; it was, Pepys said, 'the best collection of anything almost that ever I saw'. His book was similarly well liked. Pepys called it 'very pretty', and the fledgling Royal Society of Philosophers pronounced it so impressive that Graunt was elected to the Royal Society in the year of publication. His admission was approved personally by King Charles II, who was bowled over that a London shopkeeper could be capable of such insightful work. The King remarked that 'if they found any more such Tradesmen, they should be sure to admit them all, without any more ado'.

As well as a prediction of lifespan depending on age — the foundation of the life insurance industry — Graunt offered a number of innovations in his *Observations*. He looked at christening records as well as deaths, and noted that the birth rate was lower for females and the death rate was higher for males, evening things out. He made a reasonable estimate of London's population, and compiled a table comparing the city's mortality rates from different diseases. He showed that the plague did not spread directly from person to person (he didn't have the data to show that it was, in fact, spread by rat fleas), and dispensed with the superstition that plague outbreaks coincide with the coronation of a monarch. He even created 'life tables' that predicted how many people would live to what age and the averaged life expectancy of each generation. Perhaps most importantly, he knew his analysis could be unreliable, and used multiple angles of attack on a single question, providing checks for the conclusions drawn.

Despite the praise heaped upon his work, Graunt's late-life conversion to the widely reviled Catholic faith led his business into trouble, and the Graunts into poverty. He died a little over a decade later, and the Drapers Company paid his widow an annual pension of £4 'in regard of her low condition'. His book was still being reprinted, though; at his death from jaundice in 1674, Graunt became one of his own statistics.

Living by Numbers

Graunt actually beat the odds. By dying just before his 54th birthday, he had surpassed the life expectancy in England at the time by around 20 years. In the UK — the country with the longest dataset — lifespan remained between 30 and 40 years from the 1540s to the beginning of the 19th century. The story was similar everywhere: in 1800, life expectancy was no more than 40 years in any country in the world. But in 2019, the average global life expectancy was 73 years.[2] What happened? We developed effective medicines and took control of infectious disease. And we couldn't have done it without statistics.

At its heart, statistics is just a set of tools. You apply the tools to sets of numbers in order to be able to say something reasonably accurate about what those numbers describe. It doesn't sound like much, but it is an extraordinarily powerful invention. Statistics allows us to make a change to whatever we've measured, then take a new set of measurements and see whether the change made things better or worse — and to know how confident we should be in our conclusions.

This rather simple, boring set of tools has had a profound effect on humanity; no wonder the nurse and medical statistician Florence Nightingale reportedly believed that the study of statistics would reveal God's thoughts. This might be overstating the case, but those who can use statistics are certainly a powerful bunch. That's why, when he was a senior vice-president at Google, Jonathan Rosenberg declared that, 'Data is the sword of the 21st century, those who wield it well, the Samurai'.[3]

To those in the know, this sentiment is actually a double-edged sword. While we tend to think of the samurai as glamorous, skilled warriors, the truth is that many of them ended their careers as bureaucrats who collected and processed statistics — the kinds of people we think of as a little dull. It *is* a little disappointing to learn this, but the samurai were highly educated members of Japanese society, and when there was no more need for warriors, they did indeed lay down their swords and become civil servants.

Civil servants have always performed statistical functions with data: the word 'statistics' comes from the German word *Statistik*, first coined in 1749, meaning 'facts about the state'. In England, the discipline was originally known as 'political arithmetic'; the first mention of 'statistics' in English came with the 1791 publication of *The Statistical Account of Scotland*.[4] And then, in the 19th century, the subject exploded. Not, it has to be said, for the most uplifting of reasons.

For all its benefits, we cannot pretend that the invention of statistics has produced nothing but good. Indeed, statistics might reasonably be termed the most problematic discipline of mathematics. You might have opinions about the use of logarithmic tools in the production of the atomic bomb, but at least it wasn't the application that the inventor of logarithms had in mind. You might say that we should — like Tartaglia — have reservations about the use of algebra for improving our war machines. I could counter that by suggesting that war is just part of human history, and that algebra has also — as with game theory — been used to avoid it. However, when it comes to the earliest applications of statistics to eugenics — population control based on ill-formed opinions about race, intelligence, criminality, and so on — the subject has to hang its head in shame. And much of this shame stems from the work of Francis Galton.

Galton was clever, well resourced (Charles Darwin was his cousin; both men benefited from privileged upbringings), and capable of remarkable detachment from his fellow creatures. His goal, after much work in developing the science of statistics and measurement, was to breed a race of superhumans who were free from the taints of the poor, the weak, the disabled — even the ugly.

This was no secret project: he laid out the idea in an 1869 book entitled *Hereditary Genius*. Here, he proposed that science should 'check the birth rate of the Unfit and improve the race by furthering the productivity of the fit by early marriage of the best stock'. Galton created a beauty map of Britain, concluding that the nation's ugliest women — he called them the 'repellents' — resided in the Scottish city of Aberdeen. Within the pages of *Hereditary Genius* Galton coined the term 'eugenics', from the Greek for 'well born'. In a 1904 essay entitled 'Eugenics: Its Definition, Scope and Aims', Galton wrote that 'What nature does blindly, slowly, and ruthlessly, man may do providently, quickly, and kindly. As it lies within his power, so it becomes his duty to work in that direction.'[5]

Galton's views resonated with many people. A stellar array of British and American intellectuals jumped on board with the idea of creating a master race and hindering the reproduction of 'undesirables'. H.G. Wells, for instance, once declared that, 'It is in the sterilization of failures, and not in the selection of successes for breeding, that the possibility of an improvement of the human stock lies.' In a lecture at London's Eugenics Education Society, the playwright George Bernard Shaw went even further, advocating 'an extensive use of the lethal chamber'.[6] To create a better world, he declared, 'A great many people would have to be put out of existence, simply because it wastes other people's time to look after them.'

Francis Galton was the Eugenics Education Society's founding president and a lifelong member. Winston Churchill was another notable supporter (in 1910, Churchill wrote a memo to Herbert Henry Asquith, the British prime minister,[7] warning that, 'The multiplication of the feeble-minded is a very terrible danger to the race'; two years later, he was vice-president of the first International Eugenics Conference). In 1989, in a late bid to shake off the controversy surrounding the E-word, the Eugenics Education Society was renamed the Galton Institute. This wasn't the first pro-eugenics group to bear Galton's name, either. In America, his work inspired the creation of numerous eugenics societies, the most prestigious of which was the Galton Society. One of its founders, Madison Grant, was the author of a 1916 book called *The Passing of the Great Race*. This hand-wringing lament at American immigration policy argued for the development of a better society through forced sterilisation and various other measures that would ensure the emergence of a race of superhumans. It has been described as 'the most influential tract of American scientific racism'[8], but its influence was international. In the early 1930s Adolf Hitler wrote Grant a letter, thanking him for sharing his ideas and referring to the book as 'my Bible'. When the Nazis came to power in Germany, they ordered that *The Passing of the Great Race* should be reprinted.

We cannot excuse Galton and his contemporaries on the basis of the times they lived in. Galton's half-cousin Charles Darwin was also an excellent thinker, a scientific innovator and someone willing to break new ground. But he was appalled by pseudo-scientific arguments that claimed humans could be ranked by race, and was a subtle activist for the abolition of slavery.[9] The slaver, Darwin once wrote, 'has debased his Nature & violates every best instinctive feeling by making slave of his fellow black'.

Ultimately, it's hard to know what to say about Galton. His views were appalling and inexcusable. His work set the tone for many avenues of scientific inquiry, and its legacy lives on in misguided modern-day attempts to create divisions between human populations by searching for genetic explanations for racial characteristics.[10] Nonetheless, we have to admit that Galton's legacy also includes some of our most widely used mathematical techniques. He made the modern discovery of what has come to be known as the 'wisdom of crowds', for example, where multiple uninformed guesses about a quantity can be mined for a good approximate answer.

The ancient version came during the Peloponnesian War of the 5th century BC, when a Platean commander told his men to count the brick layers in an exposed, unpainted section of the wall that the Peloponnesians and Boeotians had built around their city. The commander took the most common value that his soldiers reported — the mode — and multiplied it by his estimate of the breadth of the brick to get the height of the wall. Then he built ladders long enough for them to scale the walls and escape.

In his rediscovery of the technique, Galton used the median. In 1907, he was in the coastal town of Plymouth in southern England. Wandering through the marketplace, he happened upon a competition to guess the weight of an ox at the Fat Stock and Poultry Exhibition. His attention was captured by the glorious array of numbers on display, and after the competition was over, he persuaded the organisers to give him the tickets containing all the visitors' guesses. Discarding the illegible tickets, he arranged 787 of them in order, and worked out that the middle — median — value of 1,207 pounds was within 1 per cent of the actual weight of 1,198 pounds. He wrote up the experience in a letter to *Nature*.[11]

It's worth pausing for a moment here because we can see the lurking dangers of the naive application of statistics in this paper. Galton opens with the observation, 'In these democratic days, any investigation into the trustworthiness and peculiarities of popular judgments is of interest', and declares that, 'The average competitor was probably as well fitted for making a just estimate ... as an average voter is on judging the merits of most political issues on which he votes.' In the end, he concedes that the outcome was 'creditable to the trustworthiness of a democratic judgment'. But it was a perilous comparison nonetheless. Statisticians have learned that what works well in one context rarely works well in another. Even a cursory analysis reveals that democratic processes are different from weighing oxen. They don't rely, for instance, on taking medians, or even averages of the population's votes. You can't cite success in one context as a reason for applying the same tool to another, which is why statisticians are extremely careful about their analysis, and how it is used. After all, as the old joke goes, the statistically average human has slightly less than one testicle.

For all his appalling social views, Galton was his own harshest critic when it came to getting the numbers to reveal their secrets. In his *Nature* paper about the wisdom of crowds, he took care to include an estimate of the 'probable error' of each value. And he gave us many of the statistical tools we have developed for use in the present day. He pioneered the exploration of correlations, for instance, which determine whether two variables track each others' changes in ways that can be related.[12] Galton's introduction of the idea makes the point that you can measure an array of arm and leg lengths and decide whether the two lengths are correlated. We might now use the same idea to look for correlations such as between measures of pollution levels and admissions to hospital because of breathing difficulties.

Galton also did the initial work on separating out the effects of multiple causes, inspiring his long-time collaborator Karl Pearson to create sophisticated mathematical models that track this, and eventually to invent the 'chi-squared' test. We now use this test to interpret data from medical trials, such as the best age to vaccinate children. The phrase 'regression to the mean', which describes the tendency of a series of measures to return to average values after a foray into the territory of 'outliers', was another of Galton's inventions. He originally called it 'reversion to mediocrity', but the idea is the same. This is the kind of statistical observation that tells me that I'm about to experience an improvement in some symptoms. I know that because I went to the doctor: I only go in extreme cases, and once you've reached such an extreme, there is a much higher likelihood of a movement in the opposite direction. Regression to the mean, in other words, tells me that the worst phase of an illness is the phase when I am about to get better.

Galton also invented the questionnaire, a mainstay of medical, psychological, and sociological research. He used this in another of his creations: twin studies, where biological variables are minimised, allowing researchers to probe the effects of nurture while (almost) disregarding issues of nature. This, he said, would make it 'possible to weigh in just scales the effects of Nature and Nurture, and to ascertain their respective shares in framing the disposition and intellectual ability of men'.[13] He sent out hundreds of questionnaires to the parents of twins, asking them about similarities and differences, and how life experiences had affected them.

One thing that Galton never quite understood was that numbers aren't everything to everyone. Sometimes you want people to have a visceral reaction to your findings, not an intellectual one. And since our brains struggle with numbers, that means presenting them in ways that tap into primal circuits. Just as a picture of a screaming face can cause a burst of adrenaline, data visualisation can bypass our brains' limitations and persuade us of a hard-to-access truth. If I ask you to think about the impact of the 4 trillion plastic water bottles that we have bought in the last 10 years, for instance, you're going to be less moved by the prospect of an impending environmental catastrophe than if I show you a mountain of plastic water bottles 2.4 kilometres high about to engulf Manhattan.[14] And no one executed this strategy better than the Lady with the Lamp.

The Power of Pictures

Florence Nightingale is best known for her night-time care in the British Army hospital in Scutari, Turkey, during the Crimean war. Her fame began when a report in the London *Times* newspaper on 8 February 1855 described her as 'a "ministering angel" ... and as her slender form glides quietly along each corridor every poor fellow's face softens with gratitude at the sight of her. When all the medical officers have retired for the night, and silence and darkness have settled down upon these miles of prostrate sick, she may be observed alone, with a little lamp in her hand, making her solitary rounds.'[15]

It's no wonder she stood out to the soldiers: she was the only woman allowed on the ward after 8 pm. In an attempt to safeguard the virtue of her nurses, Nightingale kept them under lock and key — and slept with the key under her pillow.

Such measures seemed necessary to Nightingale. She was appalled at the goings-on in the hospital; writing to her friend Henry Bonham Carter, she discussed a sergeant who had used his key to the military stores to spend nights there with a woman who worked in the hospital. 'The consequences were soon obvious,' Nightingale wrote, archly, about the woman's pregnancy. She tried to get the camp's commandant to discipline the sergeant, and was outraged by the response. 'I obtained no redress — not even a reprimand to the man,' she wrote.[16] But if Nightingale was a stickler for discipline and morality, she was even more of a stickler for numbers. She understood that, properly parsed, they could save lives.

Nightingale had studied mathematics from an early age. When training in France and Germany, she had collected hospital reports, statistical data, and information on how hospital sanitation and nursing systems were organised. While in Scutari, Nightingale undertook a vast recording of the numbers of patients dying there and elsewhere, and analysed the figures to show that 37.5 per cent of patients at Scutari died, while hospitals at the battlefront had only a 12.5 per cent mortality rate.

Armed with the numbers, she set out to find the reason why — and to do something about it. How? With powerful infographics.

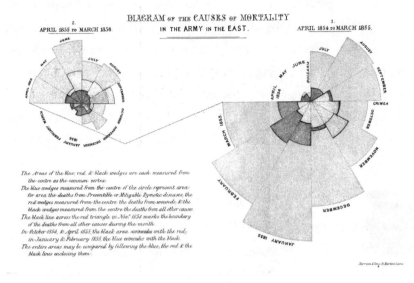

Florence Nightingale's wedge diagram

Wellcome Collection, Attribution 4.0 International (CC BY 4.0)

Nightingale's 'wedge' diagram demonstrates at a glance that disease was killing more soldiers in Crimea than battle wounds. The area of each sector represents a single month's deaths, with the cause distinguished by colour. She presented the diagram to the Secretary of State for War, then included it in her 1858 book *Notes on Matters Affecting Health, Efficiency, and Hospital Administration of the British Army*. She sent a copy to Queen Victoria, who requested that Nightingale come and explain her findings in person. As a result, she procured a Royal Commission into the health of the Army, which led to reforms in military medical practices. The diagram, she said, was vital to this: 'Diagrams are of great utility for illustrating certain questions of vital statistics by conveying ideas on the subject through the eye, which cannot be so readily grasped when constrained in figures.'

Florence Nightingale was more than a nurse, and more than a

statistician; she was a hugely effective lobbyist. After the *Times* report made her famous, she even leveraged her celebrity. There were downsides to fame — Nightingale had to sneak back into Britain in August 1856 under a false name to avoid being mobbed — but it also helped her raise over £40,000 for the 'Nightingale Fund', enough to set up the Nightingale Nurse Training School at St Thomas' Hospital in London. To cap it all, Nightingale was elected as the first female member of the Royal Statistical Society in 1859. This was no celebrity appointment: she was recognised as a *tour de force* who had been working in the field for many decades.

Searching for Significance

By the time Florence Nightingale pioneered her charts, statisticians had already pulled together a range of tools for deciphering data. The first was a method of finding the simplest curve that best describes the main trend in a set of scattered data points. It was called the 'least-squares' method, and provided a way to make a line run past each data point as closely as is possible while remaining smooth.

Mathematicians argue about who came up with the least-squares method. The Frenchman Adrien-Marie Legendre published a version in 1805, but the German Carl Friedrich Gauss published a more complete take on the problem in 1809 (a year after Robert Adrain, a secondary school teacher in the United States, published *his* equally good version). Legendre, Gauss, and Adrain all found a formula that works through 'residuals', the vertical distances to the line from each data point. Because there are points on each side of the line, some will be positive and some negative, so the first step is to square them to get rid of the negatives. The least-squares line is the one that minimises the total of the squared residuals.

Much more interesting is Gauss's 1809 'normal distribution'. A 'distribution' is a scatter of data. This can be plotted out in various ways; the normal — or Gaussian — distribution is a form where three particular properties of the data are identical. Those properties are the mean, the mode, and the median. We met two of these when looking at Francis Galton's work; together with the mode, they provide three different ways of calculating what we laypeople would term an 'average'.

Imagine a dataset composed of the heights of all the people who live in your street, for example. The mean is calculated by adding up all the heights and dividing by the number of heights you've added together. The mode is the height shared by the largest number of people. You get the median by arranging all the people in a line, from shortest to tallest, and taking the height of the person in the middle of the line. In a normal distribution, the mean, mode, and median all have the same value. It has other interesting properties concerning how the data are distributed, which we'll get to in a little while.

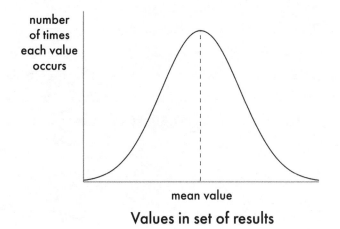

The normal distribution

Heights are just one example of values that tend to follow a normal distribution. School test results are distributed in the same way, as is blood pressure in a population. As any actuary or life insurance specialist will tell you, the ages to which a population of humans live follow a normal distribution, though this is slightly skewed (understanding exactly how it's skewed is how they make their money). The normal distribution is everywhere; though the origin of its name is not quite clear, you can certainly think of the normal distribution as being the norm for data.

The normal distribution tends to arise when a large number of independent factors each exert a very small amount of influence on the thing being measured (the various genetic, social, and evolutionary factors that determine any human's height, for example), but there are other ways in which data get distributed. There's the one discovered by Siméon-Denis Poisson, for instance.

As we saw when we came across Euler's number e in Chapter 5, Poisson distributions occur when an event occurs rarely but repeatedly, and each occurrence is independent of any other. Poisson was looking at the rate of wrongful convictions in Parisian courts in the 1820s, and whether juries had become less inclined to convict their fellow citizens (they hadn't).[17] These days, we can see Poisson distributions in various systems, such as the number of goals scored in soccer matches (in the English Premier League, 2 and 3 are most common), and in the probable number of meteorites above a certain size hitting the Earth in a year (for meteorites above 22.4 metres in diameter, 10, 11, or 12 impacts are the most probable).

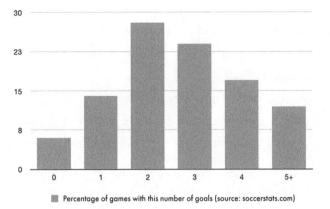

Percentage of games with this number of goals (source: soccerstats.com)

An example of a Poisson distribution: distribution of goals in English Premier League soccer matches in the 2019–20 season

In each case, it is possible to compute an average, and apply that average to Poisson statistics to make predictions. Imagine, for example, that I'm running a bar, and I know that I get through 10 cases of beer a night, on average. How do I plan for an unexpected surge in customers? It's no use buying in 20 cases to cover all eventualities, because that's too much financial outlay. If I buy in too few cases, though — say, just 12 — I risk running out and looking like I don't know how to run a bar. The new customers will be lost for ever.

It turns out that I can make an educated guess using Poisson statistics. There's a formula that gives me the probability of needing x cases. It involves the historical average number of cases needed λ and (of course — it's everywhere) e, Euler's number:

$$P(x) = \frac{e^{-\lambda}\lambda^x}{x!}$$

(The exclamation mark after the x denotes 'factorial', which means x multiplied by $x - 1$, $x - 2$, $x - 3$, and so on, all the way down to 1).

The probability (P) of needing 15 cases of beer is just 3.5 per cent. I'll use up 13 cases on just 7.3 per cent of the nights. A stock of 12 cases would cover 9.5 per cent of the evenings.

So, how many should I have in stock? If I can afford it, maybe 15 ... I'd only use all those up on (roughly) 12 nights in the whole year. But, really, it's a judgement call.

This is an important point. At heart, statistics is always about making judgement calls. It is the science of educated guesses, if you like. It looks like maths, it smells like maths, but there's none of the perfect certainty we associate with maths. It's all about saying what's likely, given certain numbers — and given several assumptions about how trustworthy those numbers are. And perhaps that's why we humans who have made the effort to learn some mathematics still struggle with what statistics has to offer.

We have seen, right from the start of this journey, that our brains aren't great with numbers. When it comes to statistics, that's especially true. We see statistical figures, and we forget the caveats that accompany them. Or we can't process exactly what they mean. If I tell you, for instance, that the World Health Organization has said that eating 50 grams of processed meat — something like a two-rasher bacon sandwich — every day increases your chances of developing bowel cancer by 18 per cent, is that a problem to you?[18]

If it sounds worrying, that might be because you haven't properly processed the word 'increases' in the previous sentence. That daily sandwich consumption does not increase your chances of getting cancer some day by 18 per cent. It increases your chance of joining the 6 per cent of people who don't eat a bacon sandwich every day but end up developing bowel cancer at some point in their life. That is where the 18 per cent increase comes in: it adds 18 per cent of 6 per cent to your chance of being among bowel cancer sufferers.

Now 18 per cent of 6 is 1.08. So instead of a 6 per cent chance, you are at 6 + 1.08, which is a 7.08 per cent chance. The chances are that you never worried about a 6 per cent chance of developing bowel cancer. Are you so concerned about a 7 per cent chance that you're not going to eat a bacon sandwich every day?

It probably wouldn't be the most rational reason for not eating that much processed meat anyway. And you probably don't eat that much in the first place. Also, given how much pleasure you can get from eating a bacon sandwich — and how doing pleasurable things is often good for your health — it's definitely a judgement call.

The same is true with the question of whether particle physicists detected the Higgs boson at CERN in Geneva in 2012. They can't be 100 per cent sure that it wasn't some fluke occurrence in the particle detector (or rather, some series of fluke occurrences, since the verdict was based on multiple events in their detectors). When we say we're certain of something in science, what we actually mean is only that it's very, very unlikely that it happened by chance.

Statisticians quantify this certainty by number-crunching various attributes of the data — its mean and the sample size, for instance. One essential component is the 'standard deviation', which is a measure of how different, on average, the various sample values are from the mean value. The units of this are the same as the units of whatever you're measuring: if it is the heights of 101 Dalmatians, the mean height might be 60cm and the standard deviation might be 3cm.

The standard deviation provides a useful perspective on the data. Assuming the dogs' heights follow a normal distribution centred around 60cm, a standard deviation of 3cm tells us that 68 per cent of the dogs' heights lie between 57cm and 63cm. In this context, this range is known as 1 standard deviation, or 1σ (sigma). Two standard deviations (2σ) is the span of heights that account for 95 per cent of the dogs' heights. And 3σ is 99.7 per cent.

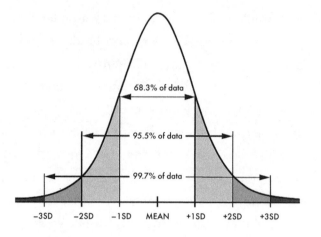

Standard deviations on a normal distribution

Going back to the Higgs boson, confidence in an experimental result comes down to a figure derived from the standard deviation and other attributes of the data. It is called the '*p*-value', and it's not straightforward to define in an easily intelligible way. Press a statistician about it, and they'll give you a word soup along the lines of 'the *p*-value is the probability of obtaining results at least as extreme as the observed results of a hypothesis test, assuming that the hypothesis is correct'. My hand-waving alternative is that the *p*-value is a measure of how likely you are to get a particular result if it *isn't* caused by what you think it is. Let me put it like this: if the result measured at CERN is *nothing* to do with the Higgs boson, but just a chance fluctuation in the background noise in the experiment, how many times would CERN scientists get that result — the result that suggests a Higgs boson *is* there — when repeating the experiment?

To decide, we have to make another judgement call. This one was first made by a British statistician called Ronald Fisher. In 1925, he published a book called *Statistical Methods for Research Workers*.[19] In its pages, he suggested that 1 in 20 was a decent cut-off for most experiments that try to determine whether an observed phenomenon has an interesting cause or is just due to chance. This is the factor known as 'statistical significance'.

Fisher's 1 in 20 is equivalent to 5 per cent. However, Fisher's cut-off is not as simple as 'if we get that result 95 per cent of the time, that's the answer'. If only. To demonstrate statistical significance actually involves negotiating a complicated series of tricky steps.

First, we have to set up a hypothesis, and test to see whether it's true. In the case of the Higgs boson, it would be related to the data that come out of the detectors, measuring (roughly) the number of times they are hit with a particular amount of energy. According to the theory, the presence of a Higgs boson will create an idiosyncratic bump in the number of detections of a particular value of energy.

Of course, there's always a chance that random fluctuations in the detectors could cause that odd pattern to pop up: fluctuations from electrical noise, say, or an array of energetic particles that just happen to hit the detectors after travelling through interstellar space. That means you can never be sure that any one particular bump is due to the Higgs boson. But you can set things up so that it's extremely unlikely that you would get a particular dataset from the apparatus unless a Higgs boson were responsible. Now you just have to decide how unlikely is unlikely enough.

The scientific gold standard for a 'discovery' used by CERN for the Higgs boson is 5σ, which corresponds to 99.99994 per cent of the recorded values lying within the range. What does that mean? Parsing the complicated statistical routines correctly tells you that, on average, we would see that same Nobel Prize–winning dataset once in every 3.5 million runs of the experiment if the Higgs boson does *not* exist. In case you're interested, that corresponds to a *p*-value of 0.0000003.

If we lowered the standard for discovery from 5σ to 1σ, we would only have to repeat the experiment 6 times (on average) to get a chance result that looks like a discovery. If the threshold were 3σ, an average of 741 experimental runs would give us the misleading 'discovery'. You could argue that, to be sure, we should insist on more than 5σ — why not 10σ? This is the judgement call. The truth is, if we required 10σ, we'd probably never be able to say we had discovered anything. And it's worth pointing out that we haven't always required 5σ. In 1984, CERN scientists Carlo Rubbia and Simon van der Meer won the Nobel Prize for physics after 'discovering' the W and Z bosons the previous year. But the statistical significance of their results was not even close to 5σ.[20]

This judgement call issue is important, because it is central to whether a life-saving drug makes it to market, or whether a defendant is found guilty in a court of law. Let's do the legal side first, because it matters far more than many of us appreciate.

Crime and Punishment

Depending on where in the world you live, there is around a 1 in 3 chance that you will be asked to serve on a jury at some point in your life. During that service, you may be asked to assess statistical evidence. You will almost certainly have no training in this. And it's entirely possible that the person presenting the evidence won't have sufficient training either. It's a serious issue in the justice system, and it has killed people.

I'm thinking in particular of Sally Clark, a British woman who was convicted in November 1999 of killing her first two babies. The defence insisted the babies had died from natural causes — either some undiagnosed familial health issue, or the heartbreaking, inexplicable 'sudden infant death syndrome' (SIDS). An expert on child abuse — a medical expert — called Professor Sir Roy Meadow told the court that the statistical risk of a child dying from SIDS in a household like Sally Clark's was 1 in 8,453. The chances of the same thing happening a second time, he said, would be 1 in 8,543 squared: 1 in 73 million. In other words, it's the longest of long shots.

The jury convicted Sally Clark. An immediate appeal was unsuccessful. A second appeal was upheld and Clark was freed on the basis that her conviction was statistically unsound. But not before another woman had been convicted of another double murder in an eerily similar case (also involving Meadows' testimony), and not before Sally Clark's mental health was so catastrophically destroyed that she drank herself to death four years after her release.

There were several problems with the prosecution of Sally Clark, but we'll focus on just two.[21, 22] First, even if the lifestyle and conditions in the Clark house did mean SIDS was a 1 in 8,453 event, it simply doesn't follow that the chance of a second SIDS death in the house is just as unlikely. You can't just square the probabilities. If some unknown factor made it happen once, there's a reasonable chance that the same unknown cause could make it happen again. In other words, another inexplicable death becomes much more likely (one estimate has put it as high as 1 in 60). Second, just because the explanation that would render the defendant innocent is highly improbable, that doesn't make guilt highly probable. This strategy of introducing spurious or misleading statistics that seem to indicate a low likelihood of guilt is known as the prosecutor's fallacy.

It's worth mentioning there is a defendant's fallacy too. It was used in the O.J. Simpson trial: the defence team raised the fact that fewer than 1 in 1,000 domestic abusers of women end up killing them. If you're on the jury, you are now asking yourself, is O.J. Simpson worse than 1 in 1,000? But you shouldn't; that's the confusing question they want to distract you with. The point is, Nicole Brown was dead, and the relevant statistic is actually this: 4 in 5 abused and murdered women are killed by their partner.

Even scientifically sound evidence such as DNA matches can be misleading if statistics are handled badly. A US jury in a robbery trial might be presented with the notion that a match between a suspect's DNA and DNA found at the crime scene is a '1 in a million' chance, for example. That might make the jury think it's an open-and-shut case. But there are 152 million adult males in the US, which means there are potentially 151 other matches besides the suspect. There has to be more evidence than just this to secure a conviction.

A similar problem has arisen in testing populations for Covid-19 in the recent viral pandemic. In an imperfect world, a test that is '99 per cent accurate' sounds close enough to perfect, doesn't it? So, we should surely use it to test everyone, whether or not they have symptoms? If they test positive, they can take time to recover (if they need to), then go about their business, secure in the knowledge that they won't get sick again because they'll have immunity. But that's potentially a very bad judgement call.

Let's say one person in every thousand actually has Covid-19, and we're testing 1,000 people. A test with 99 per cent accuracy will give the right answer to 99 per cent of people who are sick, and to 99 per cent of those who are healthy. That means it will give a positive result to 1 per cent of the remaining 999 people who don't have Covid. That's a whopping 9.99 people. Effectively, 11 people out of 1,000 will get a positive test, and only 1 of them will actually have developed immunity. So if you have a positive test, you can only be 10 per cent sure that you have immunity. It's not a great help is it? And that's with *all* the information. When you don't actually know the prevalence of Covid-19 in the population, you don't know how many false positives you're getting.

The counterintuitive nature of all this is one reason why many statisticians prefer to work with a different system. It's called Bayesian analysis, and it is not a new invention — far from it. The Reverend Thomas Bayes developed his theorem in the middle of the 18th century. We don't know exactly when, because Bayes never told anyone about his ideas.

The Bayes statistical system — described in documents found after the Reverend's death in 1761 — still causes arguments between statisticians. No one can agree whether it is better than the standard 'frequentist' system we have discussed thus far. The familiar system is called frequentist because it relies on finding a probability through an analysis of the frequency with which an outcome occurs. If I repeatedly roll a fair die, for instance, I will get all of the numbers with equal frequency in the long run. Bayes' system, on the other hand, is about looking at 'conditional probabilities': what are the chances of *B*, given that *A* has happened?

You might, for instance, be a juror who has heard evidence that makes you 70 per cent sure that I am guilty of an assault charge. But you haven't heard the forensic evidence yet. When it is presented, you discover that the blood found on the victim has the same blood group as mine. Aha! But wait: 35 per cent of the population have that blood group. Should that make you more convinced that I'm guilty, or less? Or is it irrelevant information?

With Bayesian statistics, you can sit among the jurors and work it out with a pencil and paper. The calculation is actually catastrophic for me: you should now be roughly half as convinced as before that I am innocent. That doesn't mean you're 140 per cent sure I'm guilty, by the way. The way it works is that you were 30 per cent convinced of my innocence, but the forensic evidence has strengthened your hunch about my guilt. Now you are only 14 per cent convinced of my innocence, and thus 86 per cent sure I did it.[23]

It might sound far-fetched that a juror could crunch numbers on their perception of a defendant's guilt, but be assured, it's been happening for a long time. Take the 1993 case of *New Jersey* v *Spann*, for instance.[24] Here, a black male prison officer called Joseph Spann was accused of conceiving a child with a female inmate — a crime, given his position. Everything hinged on whether the prosecution could prove Spann was the father of the child.

The state presented forensic evidence based on genetic testing that, they claimed, gave a 96.55 per cent probability of Spann being the father. This took into account the fact that the child had a particular set of genes that weren't present in the mother's DNA, but were generally present in 1 per cent of black males in America — and Spann was in that 1 per cent. The jury were told they could work from whatever 'prior' sense of his guilt they felt was appropriate — but they were also told that the state's expert witness put it at a 'neutral' 50 per cent. The ins and outs of the case actually make fascinating reading.[25] The expert testified that a resulting probability of paternity of less than 90 per cent is thought of as 'not useful', 90–94.99 per cent could be interpreted as 'likely', 95–99 per cent corresponded to 'very likely', and 99.1–99.79 per cent was 'extremely likely'.

The jury were shown how to calculate their own judgement call using their chosen priors. However, they were not coached in how their priors would change their estimations. In the end, they found the defendant guilty, but it remains a controversial case for a number of reasons — not least the reliance on the statistical calculations made by a statistically naive jury.

If your head is set spinning by all these numbers, you're far from alone. These days, with forensic evidence becoming ever more central to the justice system, the mathematics involved is becoming almost problematic. The idea that ordinary members of the public should decide the innocence or guilt of those accused of crimes is a central pillar of our societies, but it's hard to argue against the idea that statistics is best left to the professionals. That's certainly the case in the next kind of trial we're going to cross-examine: the one that makes or breaks pharmaceutical fortunes. Let's take a quick look at the statistics — using a standard, frequentist approach — of a hypothetical drug trial.

Pain, Pills, Placebos, and p-Values

In the days before we had mastered the art of more, the efficacy of medical treatments was a matter of opinion, often based on little more than hearsay or hunches. That's how, for example, the celebrated genius Isaac Newton could convince himself that a tincture of toad vomit would cure the bubonic plague. These days, we can do better. Imagine we want to test the hypothesis that a painkiller (let's make up a name: PainDown) works better than taking no drug, or a placebo drug (a pill that contains no medicine). First, we give all patients a mild electric shock and ask them to report a pain score. Then we give half of them PainDown, and the other half a placebo pill that looks and tastes identical but has no pain-suppressing powers. Now we repeat the shock, and ask for their pain scores. The chances are that we'll get three distinct normal distributions: the PainDown curve, the placebo group curve, and the original curve. How do we tell whether PainDown is worth producing?

Through 'hypothesis testing'. Essentially, we want to find how likely it is that any recorded improvement is due to chance. Let's say that the original mean pain score is 5.71 (out of 10) for 50 people. The standard deviation of this dataset is 1.97. The group given PainDown score the pain of the electric shock at 4.28 out of 10, with a standard deviation of 1.72. Those who took a placebo now score the pain at 4.80 out of 10, with a standard deviation of 1.42.

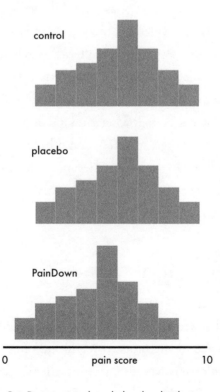

PainDown, control, and placebo distributions

We can take the numbers from the PainDown group and the placebo group and plug them into standard statistical formulae to give us a p-value. First, we create a hypothesis regarding the placebo and the PainDown results. Our 'null' hypothesis is that the difference between these groups is just down to chance. We can use a textbook statistics formula to check this and find a p value that quantifies how much trust we should put in the efficacy of PainDown. The formula takes the mean scores, standard deviation, and the number of people in the trial and gives us something called the t statistic. We convert that, using statistics software that takes into account the kind of question we're asking and the number of participants in the trial, into a p-value. In our case, the t statistic is 1.5116. That gives a p-value close to 0.072.

That's a little higher than 0.05, the usual standard for significance. Since the p-value is the probability that we would have observed a difference in mean scores as large as we did if the drug actually had no effect beyond working as a placebo, this suggests that our null hypothesis can't be rejected: this PainDown stuff is not reliably better than a placebo, it seems. The process ends, as always, with a judgement call, but this has been an enormously useful innovation. Modern medicine has been built on statistics: following this kind of routine has enabled us to quantify how well drugs, surgeries, and other medical interventions work, radically improving — and sometimes saving — countless lives, while not wasting precious health-care resources such as operating room hours and, of course, money.

Artful Extrapolation

Despite the problems we might have with the birth of statistics, and with the somewhat subjective nature of its judgement calls, its importance and influence in the modern world are undeniable. Medicine, politics, economics, justice, and science all use statistical tools in everyday practice. But there is one area of statistics that has had a far more profound influence than any other. I'm talking about sampling.

Sampling is the art of taking your knowledge of a small portion of something and reliably extrapolating that knowledge to give you a sense of the whole. You sample things all the time, but probably using intuition rather than mathematical rigour. If you're cooking spaghetti, for instance, you'll probably test more than one strand to see whether the whole pot is ready to serve. If you're looking for an electrician, you'll have a nagging feeling that you should get several quotes in order to be sure that you're not being ripped off. When you're shopping online, you make another sampling judgement call when you look at customer reviews; is the product with half a dozen five-star reviews a better bet than the one with 200 four-star reviews?

Commercial conundrums about samples are no different. If I pull out 10 stalks of ripe barley from a field and test their quality, how representative is that of the quality of the field as a whole? If I pick 10 items off a factory conveyor belt and put them through a stress test, how much confidence can I put in the remainder of the batch? Is there a way to find out if a medicine will work for most people while testing it on only a few? If I can only send only a small percentage of a signal through an electrical wire or an optical fibre, is there a way to pick the parts I send so that the person at the other end can reconstruct the original? These have become the multi-million-dollar questions that sit at the heart of our consumer-led economies.

The history of sampling can be traced back to at least AD 400. In the Sanskrit epic, the *Mahabharata*, the ancient Indian king Rituparna estimates the amount of fruit on two spreading branches of a Vibhitaka tree. He counts the fruit on a few small twigs and declares that the whole tree has 2,095 fruits and 100,000 leaves. His companion, King Nala, stays up all night to check, and confirms the number. In England, a similar process has been carried out since 1282 to check the currency produced by the Royal Mint. It is called the Trial of the Pyx, and involves sampling the new coins to check their consistency.

The name comes from the Latin for 'small box'. A random sample of each denomination of coins is picked out and locked away in a series of wooden chests. They are then taken for checks: their weight, composition, size, and markings are all compared against the desired standard. The number of coins of each denomination selected is in proportion to the number that have been struck; the idea is that, if their characteristics are within agreed limits, the entire batch will almost certainly be within them too.

The Trial of the Pyx takes place every year in the Hall of the Worshipful Company of Goldsmiths in London. The ceremony is striking in its pomp and colour, but adheres strictly to its ancient protocols. The Chancellor of the Exchequer — or a nominated representative — must be in attendance, along with the freemen of the Worshipful Company of Goldsmiths and various officers of the Mint. The decision as to whether the Mint has struck its new coins well is made by a jury composed of members of the Company of Goldsmiths. If all is well, any coins found in circulation that don't match up quite reasonably be attributed to forgers.

The Royal Mint has always been proud of its currency and ruthless in its control of forgery. In 1696 it appointed a particularly zealous man as Warden of the Mint, a position that involves, among other things, preventing forgery. Isaac Newton had never held public office before, but he brought all his powers of observation, planning, and insight to the role. By 1699 he had personally hunted down and arrested William Chaloner, the most notorious counterfeiter of them all. Chaloner was so skilled and so self-possessed that he had even raised his expertise with the government in an 'it takes a thief to catch a thief' offer to help curtail the scourge of counterfeiting. He was no match for Newton, though, and went to the gallows on 16 March.[26] Shortly afterwards, Newton received an even more lucrative job: Master of the Mint. It was in this position that he fell foul of the Trial of the Pyx.

Knowing what we do of Isaac Newton, it's not hard to imagine his unbridled rage at the 1710 jury's report that the coinage was under value by 1 part in 1,000. Newton felt that under his control, the Mint was now producing coinage with a 'much greater degree of exactness than ever was known before'.[27]

Newton's protests were so loud that the Goldsmiths threw him out of their hall. He was unrepentant, and went away and worked in laboratories with pen and paper until he found the source of their error. It was a faulty 'trial plate', the comparison standard made of gold alloy, which had been replaced that year.

Having explained the errors in the trial plate's manufacture that had led to his denouncement, Newton went on to suggest a revision to the Pyx trial protocol. The Mint, he said, should substitute the trial plate for one made of fine gold. The idea was rejected. And then, 150 years later, officials at the Mint officials acknowledged that Newton's idea was a good one. A full 133 years after his burial in Westminster Abbey, Newton's innovation was implemented. Yes, he was cantankerous, but it took more than a century for us to finish implementing his wisdom.

Guinness and the Student's *t*-Test

Statistical innovation with sampling also lies behind the success of Guinness, one of the 20th century's most successful global brands. And this one paid handsomely, too: the Guinness brewery's efforts to improve its stout gave us one of the most widely used tools in statistics.

When William Sealy Gosset joined Guinness in 1899, he was one of six men on the firm's new team of scientific brewers.[28] Each one had a first class degree in chemistry from Oxford or Cambridge, and they were treated like rock stars. They had their own accommodation in the Guinness house. Junior employees were told that if they were lucky enough to meet a brewer in the corridors, they were to lower their eyes until he had passed.

Guinness had just scaled up its operations and was determined to put science at the centre of its business. In 1886, the company had floated on the London Stock Exchange, with a great deal of success. By the time Gosset joined, it was the largest brewery in the world, which meant it needed vast quantities of hops and barley, all of a consistently high quality. The new brewers began to accumulate relevant data, but it was difficult to analyse. Despite their status and their degrees, they were uncomfortable with mathematics and almost entirely unfamiliar with statistics. As the least bad mathematician, it fell to Gosset to learn what was needed. He read a couple of textbooks, and by 1903 he could use the standard deviation and the sample size to calculate what is known as a 'standard error'. He even constructed a home-made measure of correlation. Gosset wrote a report for the brewery, outlining his new crop of statistics tools, and how they might improve production. In it, he mentioned that the reason this was new to everyone in the brewery — including the scientific brewers — was due to 'the popular dread of mathematics'. See? It's not just you.

In the summer of 1905, Guinness sent their new statistics expert to England to consult with Galton's acolyte Karl Pearson, who was now widely acknowledged as the world's leading statistician. Gosset explained that he needed to know how to make comparisons between a small number of different items; Guinness's experiments with barley, for instance, involved just four different varieties. Discerning the standard deviation in a sample of four is extremely hard to do accurately, and Gosset was hoping Pearson might have a way of at least estimating the error, and of making the relevant judgement calls, such as deciding what level of probability should be declared 'significant'. But no one, Pearson included, had the statistical tools to deal with small trials at that time. Pearson let Gosset down gently, then taught him all the statistics he knew. It took about half an hour, Gosset said.

Surprisingly, this was useful enough for Gosset to return to Guinness and implement some data analysis. And this was deemed successful enough that the brewery sent Gosset back to work with Pearson a year later, when he became a student at University College London. By 1907, with the help of what he later termed 'inspired guesses', Gosset had the answer to his questions about the errors in small samples. The research was done not using barley data but using measurements of the heights and left middle finger lengths of criminals in a local prison, with data provided by New Scotland Yard — available, as we'll see shortly, because Francis Galton had made a bid to capture (and remove) the essence of criminality from within England's population.

With the problem solved, Gosset returned once more to Dublin and implemented the new statistical laws. They made it clear that a variety called Archer provided the best barley for Guinness's purposes, and the brewery quickly bought up all the Archer seed on the market: 1,000 barrels. A year after it was sown, they had 10,000 barrels of seed to distribute to farmers, and no one else had any. Guinness had taken control of their most important raw material.

With the barley market sewn up, Guinness allowed Gosset to publish his innovation. He wasn't allowed to use his own name on the paper, in case Guinness's competitors caught on to what they were doing. He was given the option of 'Pupil' or 'Student' as a pseudonym. It is now known as 'Student's t-test'.

The t-test allows us to get a handle on the relationship between the size of our sample and the uncertainty that this size imposes on the calculation. We can then be aware of how much confidence we should have in our results. Gosset's innovation worked well for Guinness, but the truth is that no one else paid it much attention until Ronald Fisher — the man who made the decision about what constitutes statistical significance — proved the mathematics behind it and broadened the range of subjects to which it could be applied. We now use the t-test everywhere we want to compare different samples. In medical research, we use it to test the effects of antiretroviral therapies in HIV treatment. In business studies, it allows us to examine the effects of interventions such as improvements to customer service protocols. And it is still used right where it started: in agricultural research, helping us determine the efficacy of fertilisers, the relative value of crop strains, and the safety of processed products such as milk and cheese.

The Compromises of Compression

For all Fisher's innovations, it's a different kind of sampling statistics that has taken over the world in the last few decades, one that has significantly improved our quality of life — and given us everyday acronyms such as JPEG, MPEG, MP3 and HDTV. Let's spend a few moments unpacking the mathematics of data compression.

In 2019, the population of the United States received over 1 trillion audio and video files, streamed from data servers around the world. Given the capacity of the data channels that make up the internet, it simply wouldn't have been possible unless these files were 'compressed' — that is, the amount of data within them was drastically reduced from the original. And this compression would not be possible without the statistics of sampling.

When we record a piece of music, we want the recording to contain all the information necessary to reproduce what we heard in the original. That information might be encoded in the grooves of a vinyl record, the microscopic pits in the plastic of a CD, or as 0s and 1s in a digital file, but it tells whatever is going to reproduce the music exactly what frequencies of sound to produce at any moment, and how they should relate to each other in volume. That's a lot of information, even for a 3-minute pop song. But it turns out that much of the information is unnecessary.

In the early part of the 19th century, a French mathematician called Joseph Fourier showed that any continuous signal, no matter how complicated, can be reproduced as a set of sine waves of varying frequency and amplitude. For absolutely perfect reproduction, you would need an infinite set of these waves, but Fourier showed that you can make do with a finite number. The result, which involves a (relatively) simple formula and the use of complex numbers, is called a Fourier transform.

With Fourier's innovation, scientists had an entirely new tool at their disposal. A signal that varied over time could now be represented by a sweep through the frequencies of its component parts. Moving into the 'frequency domain', as it was known, allowed them to analyse and process time-varying signals in an entirely new way. The technique became central in a number of research fields, including thermodynamics, geology and — much later — quantum mechanics.

When the world began to work with digital information, a slightly different tool emerged. Fourier's transform, applied to the discrete 0s and 1s rather than a continuous analogue waveform, became a 'discrete Fourier transform'. This is the idea behind the JPEG format, constructed by the Joint Photographic Experts Group, which in 1992 approved an official standard for compressing digital image files. But a discrete Fourier transform isn't as good as it gets — as John Tukey showed in 1965.

Tukey was born in 1915, and it quickly became obvious that his brain was built for mathematics.[29] His parents saw his talent early, and educated him at home during the 1920s. He was a full professor at Princeton by the time he reached his 35th birthday, and in 1965 he founded the university's statistics department. The fast Fourier transform (FFT) was born that same year, when Tukey, a member of President Kennedy's Science Advisory Committee, realised there was a need for fast processing of seismological signals that might indicate a Russian nuclear bomb test.

Tukey, who has been described as 'a great bear of a man', had already coined the term 'bit' for binary digit used in information theory (which we'll get to in the next chapter). That was in 1947. In 1958, he invented the term 'software'. The 'fast Fourier transform' was a little less catchy, perhaps. But, as the digital revolution unfolded, it was to be just as important.

The FFT is, essentially, a shortcut for computing the discrete Fourier transform used for digital compression. JPEG didn't need the speed of the FFT. However, MPEG, the standard approved by the Moving Pictures Experts Group in 1993, certainly did. The audio layer of these moving pictures is provided by the MP3 file — the kind that you stream through wireless, bluetooth, and maybe copper or optical fibres every day. MP4 is the audio plus video. It employs the statistical analysis of Tukey's FFT to process the original signal in ways that reduce the size of the recorded data file while not noticeably reducing picture or sound quality. Its third iteration, MPEG-3, is what lights up your high-definition television. In case it's ever of interest, MPEG have already developed a standard for compressing and transmitting the information in your genome (MPEG-G) and fully immersive, 360° virtual reality video (MPEG-I). Yes, its contribution is subtle, but statistics is making life in the 21st century quite extraordinary, efficiently delivering entertainment, educational resources, business-critical data, and even personalised health care to precisely where it is needed, wherever in the world that may be.

There's just one more person we need to mention here, and she couldn't be more down to earth. Joseph Fourier was orphaned at the tender age of 9 years old, and went on to discover the greenhouse effect that drives global warming, be imprisoned during the French Revolution, and travel across continents as Napoleon Bonaparte's scientific advisor. John Tukey, as we saw, was a child genius who went on to serve President Kennedy and play a decisive role in the Cold War. I can't tell you anything particularly extraordinary about Ingrid Daubechies (unless you count the fact that in 1994 she became the first female full mathematics professor at Princeton, but that probably tells you more about Princeton than about her). Born in Belgium, and now working at Duke University in Durham, North Carolina, Daubechies is simply a very gifted mathematician who has given us the statistical tool that powers the FBI database of fingerprints, numerous life-saving medical technologies, and the instruments that routinely detect collisions of black holes a billion or so light years away. But she would hate to make a fuss about it.

Profiles of Daubechies tend to focus on her love of gardening. Maybe that's because it's not really possible for most of us to appreciate the complex mathematics of Daubechies' invention. Nonetheless, we can at least wave a hand at what are, rather cutely, called wavelets.

Wavelets provide a mathematical way to represent a blip — a very short, isolated signal, something like the spike on a heartbeat monitor. This is much harder than it sounds. When we re-create signals as collections of sine waves, they almost always have a 'long tail'; you have to use extremely high-frequency sine waves to make the signal stop suddenly. This is actually a very restricting limitation, because it adds a huge amount of data to the signal — more than was contained in the original, in many cases.

Daubechies' wavelets are an alternative to the Fourier transform system, and the wavelets occupy an infinite-dimensional space (it's not as hard as it sounds; to use them, you just invoke the power of infinite series that we encountered in the formulation of calculus). She found a way to create an original signal, known as the 'mother', that has no tail at all — the signal goes to exactly zero at a very short distance away from its peak. She then makes adjustments to the mother to create daughters, granddaughters, great-granddaughters, and so on. These provide increasing levels of detail, and can be summed together to create extremely short, information-rich blips that can be encoded in very small data files.

Daubechies made her breakthrough in 1986, and it had an immediate impact on data-handling. Medical imaging is a particularly strong application. Endoscopic, ultrasound, X-ray, MRI and CT-scan images all use wavelet processing to make the scans easy to process and transmit without losing vital — possibly life-saving — detail. But the most world-changing application of wavelets is probably in fingerprint records.

It was Francis Galton who pioneered the development of fingerprints as a resource for law enforcement. In an 1888 letter to *Nature*, he suggested they might be 'the most beautiful and characteristic of all superficial marks', describing them, rather oddly, as 'the fine lines of which the buttered fingers of children are apt to stamp impressions on the margins of the books they handle'.[30] He worked out that the chances of prints from two different people being identical were 1 in 64 billion. The opportunity was not squandered: Scotland Yard established its first fingerprint bureau in 1901, and fingerprint evidence was first presented in court within a year. The idea was adopted by New York state prisons for identifying inmates in 1903.

The value of fingerprints is proportional to the number of them you have on record, and inversely proportional to the time it takes to access and compare them. The trouble is, the more you have in your database, the longer it takes to search. Compressing fingerprints using Fourier transforms didn't help solve this paradox because a useful data compression lost too many useful details of the prints. But when wavelets came along, everything changed. Today, the FBI Criminal Justice Information Service holds the fingerprint records, encoded in Daubechies' wavelets, of around 150 million people.

Finding a Fraud

As we have seen, there is more than one way that statisticians can put people behind bars — or exonerate them. But perhaps the most extraordinary of them is Benford's law. At first glance it seems truly ridiculous. Here it is, in a nutshell: in any table of numbers that records natural activity — including the activities of humans — there will be a particular pattern in the numbers, where the digit 1 is the most frequent, followed by 2, then 3, and so on down to 9, which will occur only 4.6 per cent of the time.

The astronomer Simon Newcomb was the first to spot it, when

he analysed how his 19th-century contemporaries were using books containing tables of logarithms.[31] He found the pages were dirtier at the beginning of the book, where people had been looking up numbers that began with a 1. The pages got gradually cleaner, the further you went through the book. The tables for numbers beginning with 9 were hardly used. Newcomb concluded that most of his colleagues worked on problems that used lower digits more than higher ones. As it turns out, astronomy is only one tiny sector of the world where this is true.

This universal truth is now named after the physicist and engineer Frank Benford. In 1889, when Benford was just six, the South Fork Dam above his home town of Johnstown, Pennsylvania, suffered a catastrophic failure. The surge of floodwater hit Johnstown at 40 miles per hour, killing 2,200 people. Benford suffered a broken arm but survived: he spent the night of the flood clinging to driftwood with his one good arm.[32]

The economics of life in post-flood Johnstown forced Benford to leave school at 12, but he returned to education later, entering the University of Michigan at 23. On graduation, he joined the General Electric company, working in the Illumination Engineering laboratories in Schenectady. He worked for them for 38 years — no doubt coming into the orbit of Charles Proteus Steinmetz — and retired in July 1948. Benford died just 5 months later.

His name survives because in 1938, Benford used 20,000 observations of natural phenomena, such as the area of river basins, populations of cities, molecular weights of various chemicals, and even numbers in citizens' addresses, to identify the pattern. It took until 1995 for us to understand why this pattern exists: it results from the way different distributions such as the normal and Poisson distributions occur in nature. Benford's law — Benford actually called it 'the law of anomalous numbers' — describes how these various distributions will crop up.[33] Such is our confidence in the pattern that the US Internal Revenue Service now uses Benford's law to audit company accounts for fraud. If you're ever going to falsify anything with numbers in, make sure that you've taken Benford's law into account, because statistics will always find you out.

In fact, while we can think of statistics as the science of making judgement calls, that does rather undersell its power. There can be no denying that those judgement calls have been made badly from time to time, and — as the discipline was being formed — for questionable purposes. However, they have also given us bathroom cupboards full of medicines that we know will work, access to innumerable reliable scientific discoveries, and the tools to sift evidence in court and present the truth in persuasive ways — not to mention those perfect pints of Guinness.

It's extraordinary that we have come from ancient Babylonian tax-collecting tools in Chapter 1, and ended up back looking at an American tax-checking tool. But if I have inadvertently given you with the impression that maths is just for the boring bits of life, our next (and final) chapter should put that right. There's a trip to the outer reaches of the solar system, some juggling robots, a peek into the world of espionage, and a machine that is designed to do precisely nothing. Information theory may not sound like much, but it's actually quite fun. What's more, it has nothing to do with taxes.

Chapter 8

INFORMATION THEORY

How we created the modern era

In many ways, we have come full circle. Though this chapter has a somewhat unfamiliar title, it is actually, like the first chapter, just about the power of raw numbers. In fact, it is about numbers reduced to their absolute essence — 0 and 1, the only two numbers you need in order to express all the others. It is also about the insight humans have sought within them. The binary system has given us the information age of computing, digital data, encryption, and the internet. But it is also thought to contain our best hope of finally understanding the cosmos.

The men who gave us calculus were both irredeemable mystics. Isaac Newton thought that the Bible contained encrypted secrets and spent much of his time and energy trying to decrypt them. Gottfried Leibniz, on the other hand, was obsessed by a belief that indivisible 'simple substances' are the ultimate sources of appetite, action, and perception. Every such 'monad', he said, 'makes the perceptions or expressions of external things occur in the soul at a given time, in virtue of its own laws, as if in a world apart, and as if there existed only God and itself'.[1]

Leibniz's philosophy, known as monadology, was difficult and obscure. It made him the butt of jokes. Voltaire once wrote, 'Can you really believe that a drop of urine is an infinity of monads, and that each of these has ideas, however obscure, of the universe as a whole?' Nonetheless, monadology gave Leibniz a huge appetite for examining other philosophies and seeing if he could tease out some deep truths from within them. That's why, in 1679, he wrote a manuscript that outlined the potential for building a mechanical calculating machine based on a binary number system: 0 and 1 were the only building blocks it needed.[2] It was so exciting to him because he believed that using binary numbers would open up hitherto impossible calculations, the 'alphabet of human thought', and perhaps even reveal the nature of those 'simple substances' that underlie the very foundations of reality. The fact that you could compose all numbers from just 0 and 1 suggested to him that this could be how God created the cosmos out of nothing. He wrote to his friend Joachim Bouvet, a Jesuit missionary in China, about the idea. Bouvet replied suggesting that the Chinese might have got there first — with the I-Ching.[3]

According to legend, the I-Ching, sometimes known as the Book of Changes, was based on the work of Fu Xi, a dragon with a human face. Fu Xi studied all the patterns of the universe and its contents — the constellations, the shapes of lichens on rocks, the markings on a dove's feathers, and so on — and reduced them to pictograms known as 'trigrams'. Each of these is unique, and composed of three lines. The lines are 'binary', which means they can take one of two forms: solid or broken. That gives eight possible trigrams, each representing a form, place, or phenomenon.

The eight binary trigrams of Fu Xi

Fu Xi derived all aspects of civilisation from these eight trigrams (four of them feature on the national flag of South Korea). They gave him insight into war, leadership, marriage, business, agriculture, travel, and every other activity of the human species. Around 1050 BC, Emperor Wen, the founder of the Chinese Zhou dynasty, expanded Fu Xi's discovery by doubling the trigrams into hexagrams. Their six lines gave a possible 64 combinations, which Zhou and his heirs interpreted for the people. Over the next two hundred years, these 'Zhou Changes' became a near-sacred text used for divination and advice. The binary outcomes of tossing six coins, for instance, would lead the initiated to a hexagram that should be interpreted for insight into a particular situation. Another three hundred years after that, the philosopher Confucius wrote his famous commentaries on the Changes, explaining the ethics of the system. Eventually, this rich seam of wisdom was collated into the I-Ching. Within its pages, the name and number of each hexagram is accompanied by interpretations of its significance for every circumstance: this volume of ancient wisdom contains advice for day-to-day life, a guide to the physical universe, a manifesto of ethical principles and foretellings of your own future.

Bouvet sent Leibniz a woodcut depicting the Chinese system, and a description of all its supposed powers. Leibniz immediately set to work on an article entitled 'Explanation of the binary arithmetic, which uses only the characters 1 and 0, with some remarks on its usefulness, and on the light it throws on the ancient Chinese figures of Fu Xi'. It was published, in French, in 1705.[4]

To Leibniz's dismay, almost no one cared. Even worse, during his lifetime, neither monadism nor binary arithmetic threw any light on the mysteries of the human condition, and how it fits within the grand schemes of the cosmos. A century and a half later, mathematics teacher George Boole encountered the same disappointment. Both of them would be delighted to learn that today the binary power of 0 and 1 has finally taken over the world. The information age, made manifest in digital communications, and reaching its zenith with the creation of the internet — the 20th century's own book of changes — was built on Leibniz's binary arithmetic and George Boole's laws of logic. I surely don't need to tell you how deeply this particular invention has influenced human civilisation.

The Maths of TRUE and FALSE

George Boole was not particularly good at mathematics. In 1831, aged 16, he switched positions in the classroom, abandoning his own education in order to earn money giving others a general education. He was good enough at this that he opened a school of his own in Lincoln, in England's East Midlands, just three years later. But it was a mystical experience aged 17 that would define his legacy.

After Boole's death at the age of 49, his wife Mary Everest (a niece of the surveyor who had led the British East India Company's Great Trigonometric Survey of the Indian subcontinent and given his name to the highest mountain ever surveyed) described her husband's moment of enlightenment: 'a thought struck him suddenly,' she wrote, 'a flash of psychological insight into the conditions under which a mind most readily accumulates knowledge'.[5] From this moment, teaching became little more than Boole's means of supporting himself while he obsessively researched the functioning of the mind. He became convinced that humans receive knowledge directly from something he termed 'The Unseen'. He toyed with training as an Anglican priest to give himself the chance to explore this further, but quickly decided his insights went far beyond organised religion. In fact, Boole felt he could not even capture them with words. He went back to his books and taught himself algebra and calculus so that he could work in the cosmic language of numbers.

In the end, Boole went further than his books could take him. He developed his own system of algebra — binary algebra — and was delighted to discover, later, that Leibniz had done exactly the same, and for the same reasons. Both were obsessed by reducing things to the smallest possible units in order to answer the biggest questions. But Boole dug deeper than Leibniz ever had. By the time he was done, his system could deconstruct complex reasoning into a series of statements that were simply true or false and describe how the processes of logical thought were built up from these statements.

This construction runs according to three operations that we know now as AND, OR, and NOT. The first two operations have two inputs, each of which can be TRUE or FALSE (or, as Boole recognised, 1 and 0). AND outputs a TRUE only if both its inputs are TRUE. OR outputs a TRUE if either of the inputs is TRUE, or if they are both TRUE. NOT has just one input, and gives a TRUE if its input is FALSE, and vice versa.

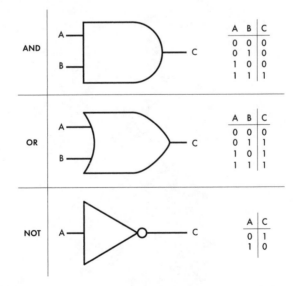

George Boole's logic operations, shown as circuit diagrams and 'truth tables' of the various outcomes

Boole published his work in 1854 as *An investigation into the Laws of Thought, on Which are founded the Mathematical Theories of Logic and Probabilities.*[6] He was so pleased with it that he hoped that it would be the thing for which he was remembered — and it was. It served, for instance, as the basis on which John Venn produced a new type of diagram in 1880.[7] Venn called them Eulerian circles, but you know them as Venn diagrams, and you can use them as a visual representation of AND, OR, and NOT. It also became the basis of modern computing. Take something like an Intel chip out of your computer, put it under a hugely powerful microscope and zoom in, and you'll eventually be looking at transistors — switches, essentially — built using electrical circuits known as logic 'gates' because they control the flow of electrical current. These gates are performing Boole's AND, OR, and NOT logic operations. There are a few ways to combine the gates into useful shortcuts, such as EXCLUSIVE-OR (XOR) gates (these give an output of TRUE only if the two inputs are different) and NOT-AND (NAND) gates, which give a TRUE in all cases except where both inputs are TRUE.

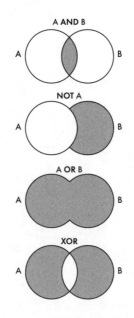

Boolean logic as Venn diagrams

Though it might not seem like much of a great leap forward now, Boole's laws of thought represented a radical new approach to mathematics, encoding what had previously been unencodable. They gained him numerous honorary degrees and a fellowship of the Royal Society. But he wasn't to enjoy this success for long; he died just ten years after the publication of his investigation.

Tragically, it was almost certainly his wife who killed him. Not out of any malice; their marriage was stupendously happy, and they felt themselves to be kindred spirits despite a 17-year age gap. Unfortunately, however, Mary was an advocate of homeopathy: she believed that like cures like. One day in November 1864, George came home soaked through and shivering after being caught in a rainstorm. Mary put him to bed and poured buckets of cold water over the sheets. He developed a cold, and then pneumonia. A few days later, he died.

Despite the accolades heaped upon Boole, it took 73 years for anyone to fully appreciate the potential of his new algebra of logic. The person to do that was — improbably — a juggling, unicycle-riding engineer called Claude Elwood Shannon.

Telephone Numbers

George Boole's vindication began with another supremely influential master's thesis. Yes, Bill Hewlett's thesis founded Silicon Valley, but Shannon's, written in 1937 and entitled 'A Symbolic Analysis of Relay and Switching Circuits', created the entire information age.[8] It sprang from his first job after gaining his undergraduate degree in electrical engineering and mathematics at the University of Michigan; he was employed at the Massachusetts Institute of Technology (MIT) to set up differential equations on a 'differential analyser', an early mechanical computer. The machine was sited in MIT's engineering department, and required Shannon to set the states of more than 100 electromechanical switches known as relays. These same relays — thousands upon thousands of them — were also the basis of the new telecommunications industry, and in 1937 Shannon's experience at MIT scored him a summer job at the Bell Laboratories of the American Telephone and Telegraph Company (AT&T). Here, he immersed himself in finding a new path through the tedious and time-consuming task of designing and testing the vast networks of relay circuits required for America's fast-growing telephone system. This was what led Shannon to rediscover the work of George Boole.

Reconfiguring the on/off states of the relays as 1/0, Shannon was able to take Boole's innovations and use them to develop binary mathematics that would represent the entire switching network of the telephone system. Instead of building, testing and then retesting thousands of switches, engineers could use Boolean algebra to write down the configurations and calculate how well they would work when built.

The benefits were immediately obvious, and Shannon received a prestigious prize for the report he wrote about the scheme.[9] When he submitted the idea for presentation at a conference being put together by the American Association of Electrical Engineers, the organisers wrote to his mentor and acclaimed Shannon's work as 'outstanding'. That mentor was Vannevar Bush, Dean of Engineering at MIT and the builder of the differential analyser. By June 1940, with Europe at war, Bush had built something new: America's National Defense Research Committee (NDRC). Many of the military research contracts he doled out went to Bell Labs — where Shannon was now working full time.

One of Shannon's first tasks for the NDRC was to help build the 'X System', a secure telephone line that enabled US President Franklin D. Roosevelt to talk confidentially with British Prime Minister Winston Churchill. Engineers see a transatlantic telephone call as nothing more than a varying electromagnetic wave, and Bell Labs' engineers knew that they could scramble the conversation if they mixed it with a series of other waves that were known only to those on either end. The sender could add the signals, and the receiver could subtract them, revealing the original transmission. However, the engineers soon realised that the mathematics of adding continuous waves made it difficult to create a fully scrambled signal; a sufficiently ingenious eavesdropper could always extract the conversation. They solved the problem by going digital.

First they broke the signal up into a series of discrete units, each one labelled by a number describing the wave's amplitude at that moment. This allowed them to add random numbers that were known only to the sender and receiver. Now it was mathematically impossible for an eavesdropper to gain any information.

Through his contributions to the X System, Shannon developed a fascination with encryption technologies. He even had discussions on the subject with the British mathematician Alan Turing, who had helped crack the German Enigma cipher, and visited Bell Labs in 1943 to learn about American innovations in encryption. As it turned out, Turing and Shannon didn't see eye to eye about the best way to encode messages, and so for most of their conversations, they steered clear of the official business, instead taking tea together to discuss what computers might be capable of. They agreed on the idea that computers could, in theory, simulate the human brain, something they thought might take a couple of decades to implement in a practical system. Evidently, the idea stuck in Turing's brain because after the war, in 1948, he wrote a groundbreaking paper entitled 'Intelligent Machinery'.[10] Shannon, though, had been distracted. His groundbreaking paper of 1948 was called 'A Mathematical Theory of Communication'.[11] Constructed on the skeleton of his extraordinary master's thesis, it is a fully formed explanation of everything that would happen in communications technology for the next 70 years.

The Birth of the Bit

The first element of Shannon's paper is the idea that information can be modelled in a development of statistical thinking. He pointed out that some combinations of words are more likely; for example, you don't expect me to follow the word 'table' with the word 'funk'. We have embodied this in the clever (but far from infallible) technology of predictive text on our phones, but it was Shannon who first demonstrated that this provides an opportunity to communicate more efficiently. Essentially, it allows many forms of communication to be 'compressed'. We can decide, for instance, not to transmit certain parts of the information because the human receiver could recover them without help. The English language is very well suited for this; its vowels are often redundant. As Shannon pointed out in an article for the *Encyclopaedia Britannica*: 'MST PPL HV LTTL DFFCLTY N RDNG THS SNTNC'.

The second element is the idea of 'information entropy'. Shannon had hung on to the potential of digitising a signal so that it could be reduced to a series of manipulable numbers. He had worked out how to quantify the information carried in a signal, too, something that had also interested Turing. Turing had called his measure of information a 'ban', but Shannon went with a suggestion from a colleague, made over a brainstorming lunch in late 1946. A binary digit can't be a 'ban', a 'bigit' or a 'binit', John Tukey said. 'Isn't the word obviously *bit*?'[12]

But how do you decide how many bits you have? Here, Shannon started with the largely uncelebrated work of an engineer called Ralph Hartley. Hartley had been working on telegraphs and voice transmission for the Western Electric Company for more than a decade when, in 1928, he published a remarkable paper called 'Transmission of Information'.[13] He had worked out that information could be quantified — regardless of the language or the transmission technology in question — through understanding the choices that lay behind it. If you toss a coin, you make one choice. If you talk to someone in English, you are making multiple choices from among the words of the English language. If you write a three-letter word in English, you make three choices from 26 possibilities. Understand the range of possible choices, Hartley said, and you get a measure of the information required to communicate. But, he added, it's not a direct measure. Choosing those three letters from the alphabet involves choosing from 17,576 (that is, 26 × 26 × 26) possibilities. Hartley noted that there wasn't *that* much information in a three-letter word, though. Instead, he suggested defining the amount of information — the number of binary (yes/no) choices to be made — via the *logarithm* (in base 2) of the total number of possibilities.

The logarithm in base 2 of 17,576 is 14.1. This means that transmitting a three-letter word in English involves making no more than 15 binary choices. The message size is 15 bits, in other words.

We can start to see the relationship when we look at bits. With one bit, you can define two possibilities: 0 or 1. With two bits, you can have four possibilities: 00, 01, 10, 11. With three bits, you can have eight possibilities: 000, 001, 010, 011, 100, 101, 110, 111.

With four bits, there are 16 possibilities. Another way to look at this is to turn it on its head and say that deciding on one out of 16 equally likely messages uses up four bits of information. There is a logarithmic relationship here: 4 is the logarithm, in base 2, of 16.

In general terms, when there are C possible choices that are equally likely, the probability of any one message being chosen is $1/C$. And the information involved in making that choice is the logarithm (base 2) of $1/C$. If some messages (or words in a language) are more likely to be used than others, the formula is a bit more complicated. First you multiply the probability of the first choice by -1, then multiply the result by the logarithm of that probability. Then you do the same with the second choice, and so on. When you have exhausted all the possibilities, you add them together to give the information content — the Shannon entropy.

To illustrate this, let's go back to our example of flipping a fair coin. Each outcome, heads or tails, has an equal probability: 1 in 2, or 0.5. The base-2 logarithm of 0.5 is −1. For the 'heads' choice, we multiply this by 0.5 and by −1. Then we do the same for tails. We add the two results. That gives 1 bit as the Shannon entropy: the amount of information conveyed in a coin toss.

In a different part of his paper, Shannon picked up on another of Hartley's observations: that the channel you are using for communication matters. If it's possible to use a wide range of frequencies in that channel — if it's 'broadband', for example — you can fit in more detail, and thus you have more choices, and can therefore send more information in a given amount of time. From this, Shannon created the mathematics of 'channel capacity'. He showed that you can characterise whatever channel you are using to transmit information in terms of the maximum number of bits you can reliably push into it (and read from it) every second. To give an (albeit simplified) example, the capacity C depends on S, the power of the signal; N, the power injected by uncontrolled, problematic noise; and W, the range of signal frequencies that the channel can handle (this is known as the bandwidth). Here's the mathematical relationship:

$$C = W \log_2 \left(1 + \frac{S}{N}\right)$$

Channel capacity is measured in bits per second — or, hopefully, in millions of bits (megabits) per second if you're measuring the capacity of your internet connection. That's why broadband internet is better than the old dial-up modems: it creates a wider bandwidth and pushes up W in the above equation. If you're far from the source of data, the signal power S is smaller, pushing C down, perhaps to the point where it takes so much time to transfer the data that you experience buffering. And if you've got a lot of interference on your internet line, N will be big, pushing C down even further. For most of us, this is our daily experience of surfing the web on phones, tablets and computers; Shannon capacity is possibly more personally relevant to us now than it was to any previous generation.

Shannon's fourth great contribution provides a way to deal with the errors that crop up when you transmit data. Every time you send a signal, there is the possibility that noise — random external factors — will compromise the receiver's ability to read and reconstruct those bits.

This isn't just a hi-tech issue: imagine you were communicating by using a light-pulse code in Roman times — maybe by moving a mirror to repeatedly reflect the Sun's rays towards your ally's army on the other hill. Someone's shield might also catch the sun, delivering a false bit. Today, if you're sending the HTML code for a web page along a copper wire, a stray electrical signal from a lightning strike or a random fluctuation in one of the circuit components can do the same. Equally, the optical fibre delivering a series of digital light pulses for a TV signal might leak some of the photons — the packets of light energy — that carry vital bits. But don't worry, Shannon's got you covered.

You might think there's an obvious solution without Shannon getting involved: just send the signal twice. It's certainly not a terrible idea. If the two signals match, you can be fairly sure you've got the right message, because a random interjection (or loss) is unlikely to happen the same way twice. But you've slowed everything down, and you've expended extra energy. And that, Shannon demonstrated, might not be necessary.

His paper includes a section on 'channel coding' where, if you know the kind of noise you will encounter, you can design an encoding system for your message that will give you mathematically perfect communication. Let's imagine, for instance, that I want to transmit four pieces of information. I can distinguish between them if I encode them as 'codewords' made of pairs of binary digits:

A	B	C	D
00	01	10	11

If I send these through a noisy channel that will occasionally flip a bit, I will risk the receiver seeing a B where I meant A, or D when I meant C. What if I send each one twice? Then C would be 1010, but with a single bit-flip of noise, that could come across as 1011. Now it looks like I've sent either C or D — it's impossible to tell which.

I could send them three times. Now C is 101010, and a single bit-flip will give 111010, 100010 or 001010. However that single bit flips, I have a 2 to 1 majority vote that says the bit is intended to be 10.

But, Shannon said, we can do better: we need only send a rather counterintuitive five bits — as long as we choose the shape of the codewords carefully so as to maximise the 'distance' between them. In this example, we'd need to encode as follows:

A	B	C	D
00000	00111	11100	11011

Here, randomly flipping any individual bit doesn't make the intention ambiguous. Try it! Astonishingly — and everyone was astonished — Shannon showed that it is possible to find optimal coding patterns for any noisy channel. In other words, there is always an error-correcting code that will allow you to transmit data at near the channel capacity (these days it's often known as the Shannon limit for error-free communication). Unfortunately, Shannon's argument was based on probability. He offered no way to show what coding patterns would make it possible to reach the Shannon limit in any given situation.

The choice of error-correction codes was pretty much the only thing that Shannon's 1948 paper left for anyone else to work on. In every other respect, information theory emerged fully formed. The only significant change made to the original 1948 paper, entitled 'A Mathematical Theory of Communication', was that it was reissued the following year as '*The* Mathematical Theory of Communication'. How's that for impact?

With Shannon's paper published, it was simply a case of waiting for the hardware that could make it all work. Within a few decades, when the required technology was sufficiently advanced, Shannon's innovation gave us email, the internet, streaming services for audio and video, data storage, and almost every other facet of what we take for granted in the 21st century. But frequently forgotten are the cultural shifts it enabled. Shannon didn't just give us video on demand. He gave us satellites, the space programme, the discovery of worlds beyond Earth, a human walking on the Moon. And all of those achievements should perhaps pale beside the biggest consequence of information theory: the discovery that Earth is a fragile, beautiful cradle for humanity that deserves protection.

On Christmas Eve 1968, the crew of *Apollo 8* beamed back the first images of our planet in a live broadcast from lunar orbit, during which they took turns reading from the book of Genesis. 'The vast loneliness is awe-inspiring and it makes you realise just what you have back there on Earth,' command module pilot Jim Lovell said.[14] It was at this moment that William Anders took the famous Earthrise photo, which has been credited with kickstarting the environmental movement. As Anders put it later, 'We set out to explore the moon and instead discovered the Earth.'[15] And they did it using Claude Shannon's theory of information.

The 'Earthrise' photo

Bill Anders, Public domain, via Wikimedia Commons

Shannon, Apollo, and the Discovery of Earth

Really, it started with *Sputnik*. In October 1957, the Soviet Union launched the first satellite, and Americans suddenly felt inferior. The following year, NASA was established. Its aim was simple: to make America the world leader in space exploration. Shortly after that, on 25 May 1961, President John F. Kennedy told Congress, 'this nation should commit itself to achieving the goal, before this decade is out, of landing a man on the moon and returning him safely to the Earth'.[16]

NASA engineers already knew that Shannon's work would be essential to this effort. Any exploration of space would require navigation, imaging, and communication signals to be exchanged between a spacecraft and Earth. Beaming signals down to Earth from space required bulky, heavy power generators, and anything that could reduce launch weight, such as manipulating the mathematics of communication theory, was a high priority.

Just a few weeks before JFK's 1961 speech, NASA's Jet Propulsion Laboratory had published a basic primer on Shannon's information theory for anyone needing to exchange signals between Earth and deep space. It was called 'Coding Theory and Its Applications to Communications Systems'.[17] It's quite fascinating to skim through it, given that we've just learned the basics of Shannon's theory. It really does read as if NASA engineers also needed to start from scratch. The second paragraph begins, 'In recent years there has been increasing emphasis on so-called *digital communications*. For purposes of this report, the digital signal may be regarded conceptually as a sequence of *ones* and *zeros* or of *ones* and *minus ones*.'

It's not very challenging stuff. The paper discusses various ways in which these binary digits can be put into 'code words' that will minimise errors in transmission. The authors demonstrate how to compute the likely errors, and then they show the results they achieved on their state-of-the-art IBM 704 computer.

Shannon is only mentioned at the end: his 1948 paper is the very last reference. It comes after the phrase, on page 75, 'Another important measure of information theory is the celebrated channel capacity.' The authors declare that, 'In the limit of infinite coding, the channel capacity can be achieved.' However, we can see how influential the 1948 paper really was from the fact that it crops up as a reference in innumerable technical documents from the Apollo programme. It also shaped NASA's 1967 budget request. Among other things, this says:

> Recurring equipment items which are necessary to maintain and update existing systems will also be required in fiscal year 1967. These include signal generators, modulating and demodulating devices, high frequency radio data modems, data quality monitors, data detection and error-correction equipment, and distortion measuring units.[18]

Data detection, error correction, and distortion measuring are the essential components of putting Shannon's theory to work in order to put a man on the Moon. Why? Because NASA had decided that Kennedy's Apollo programme would do best with all the transmissions to and from the moon — astronauts' voices; data about the spacecraft's condition, position and status; television signals; scientific results, and so on — sent through a single system. This was to be the 'Unified S-Band (USB) Transponder', and in 1963, Motorola's Government Electronics Division got the contract to build it.[19]

It was a heavy responsibility; everything in the Apollo programme was dependent on its performance. Once on the Moon, the USB Transponder was the only link to Mission Control, transmitting all the communications that we now associate with that epochal moment: those famous words 'One small step ...'; the television signal; the fuel situation; information about the landing site. If it hadn't worked, we probably wouldn't still instinctively know the names of the first men to walk on the Moon. And all this relied on implementing Shannon's information theory.

I wish I could tell you the precise details of how Shannon's work was used to shape the USB transponder system. According to a 1972 review of the system's performance, there was a specific mathematical model for the USB system. It was described in a paper entitled 'Design Philosophy of Modulation Indices for Apollo Unified S-Band Modes with Ranging'. This was written in 1965 by J.D. Hill of Bellcomm Inc. — another Bell Labs worker. Unfortunately, NASA won't let me look at Hill's paper. It is, I was told, 'not classified, but it is restricted to NASA Personnel Only and cannot be released to the public at this time'.[20] Clearly, Shannon's work, encoded in the mathematical model for the USB system, is still hot stuff.

Shannon's influence on space travel wasn't confined to the Apollo programme. In 1949, the year Shannon's theory became 'The Mathematical Theory of Communication', Marcel Golay, a Swiss-born mathematician, invented perhaps the first truly useful error-correcting code. Golay had worked in Bell Labs for four years before joining the US Army Signal Corps. Here, he rose to the position of chief scientist. His error-correction paper has just one reference — Shannon's paper of the previous year — and sets out how a group of bits can be received, error-free, through a channel that corrupts one-quarter of the bits passing through it. To run the code, you just had to transmit double the number of the original bits. The Golay code (to be precise, this formulation is known as the extended Golay code) doesn't allow you to transmit anywhere near the Shannon limit, but it was better than anything else anyone knew of — and would be for a long time to come.[21] That's why it was used aboard NASA's *Voyager 1* probe, which was launched in 1977.

The celebrated pictures of Jupiter and Saturn that *Voyager 1* beamed back were so clear and uncorrupted precisely because of Golay's developments of Shannon's work. The same craft also gave us the famous 'Pale Blue Dot' image of Earth, taken from around 6 billion kilometres away at the suggestion of astronomer Carl Sagan. In 1994, Sagan reflected on its significance. 'Look again at that dot,' he wrote.

'That's here. That's home. That's us. On it, everyone you love, everyone you know, everyone you ever heard of, every human being who ever was, lived out their lives.'[22]

Voyager's images of Jupiter and Saturn
NASA

That evocative image of Earth was taken from a distance of 6 billion kilometres, and our planet filled just a single pixel; there wasn't much scope for enhancing it. But *Voyager*'s other images could actually have been better. What nobody — or almost nobody — had realised was that another former Bell Labs researcher (and former Signal Corps engineer) had invented a better error-correcting code in 1960, long before *Voyager* had even been imagined.

To 5G and Beyond

Robert G. Gallager's progress in the Army Signal Corps was less smooth than Golay's.[23] He had been drafted into the Army from Bell Labs, and put in the 'Scientific and Professional Personnel Unit'. The unit commanders were tasked with using their troops — mostly people from Bell Labs, the Atomic Energy Commission, and graduate schools from across the US — to improve 'battlefield surveillance'. However, they squandered the brains at their disposal. Gallagher recalls training exercises where a colonel would sit in a van, write a note, and hand it to one of the scientists. The scientist then had to run to another van to deliver the note — by hand — to a different officer. If this was battlefield surveillance training, Gallager was having none of it. He wrote to his senator to report that the Army's scientific resources were being wasted. Clearly, the senator reported him: for the next three months, Gallager was reassigned to the drudge position of guard duty at the stockade. He was delighted, though: 'I had nothing to do and spent the time studying lots of things and thinking through problems,' he once recalled. 'It was a far more academic environment than anything I have experienced since.'

That can't quite be true, because when he left the Army he got a job at MIT. This was where he came up with the idea of correcting errors through a 'low-density parity check'.[24] In this scheme, the bits carrying the data are accompanied by 'parity' bits that have a protective role, almost like a 'this way up' notice on the outside of a removals box. If you find the box upside down, you check the contents haven't been damaged. Similarly, if the parity bits are flipped, the data bits have to be checked over too. It's complicated — so complicated that no one had the computing power to implement it at the time — but it allows transmissions to reach within a whisker of the Shannon limit.

Gallager's invention went unused and was eventually forgotten. But then, in 1993, two French telecoms researchers published an idea they called 'turbocodes'. These followed a similar scheme and achieved similar results to Gallager's parity checks. So similar, in fact, that, in 1996, it jogged two researchers' memories. Radford Neal and David Mackay dug out Gallager's thesis, and discovered that not only were low-density parity checks now viable, the patent on them had also run out. Why pay to license the French turbocodes when you could use Gallager's invention for free? That's certainly what numerous design engineers thought — including those who rolled out wifi standard 802.11, numerous satellite-based TV broadcasts, and the video-calling software Skype.[25]

To be fair, some users did pay for turbocodes. They were employed in 3G and 4G mobile phone communications, for instance, and to safeguard data beamed back from NASA's *Mars Reconnaissance Orbiter* (*MRO*), which launched in 2005 and is still performing great feats of communication. Indeed, the *MRO* is set to become the first link in what NASA calls an 'interplanetary Internet' that will relay signals from numerous international spacecraft as they venture further into the solar system.[26] The Deep Space Network, a network of radio stations that is vital to communications with many of NASA's interplanetary spacecraft, also uses turbocodes. It might seem prosaic now that we can communicate at rates close to the Shannon limit, but when the theory behind turbocodes were first announced, no one believed it was possible. It was only when sceptics tried them out, and failed to make them fail, that they were taken seriously.

That said, the scepticism was justified. Mathematically speaking, there was no proof turbocodes would work. Like Gallagher's parity-check code, they were an engineer's solution: a set of instructions to follow, but without explanation for why they would work. So, although both operated near the Shannon limit, they didn't operate in a way that impressed mathematicians. That was not true, however, of the 'polar codes' that sprang from the brain of Erdal Arıkan.

Arıkan is a Turkish professor in electrical and electronics engineering. In 2008, he was working on an algorithm for decoding information and saw that the techniques he was using might also be used to reach the Shannon limit. It took him two years to sort out the details, but they are now part of the latest protocol for encoding mobile phone signals in digital networks. This fifth generation (5G) protocol is known as the 5G 'New Radio' data standard. Rather satisfyingly, it works alongside Gallager's 1960 parity checks, which are also part of the 5G data transmission standard.[27] 5G is a wondrous thing: Boomer and Zoomer mathematics working side by side.

Shannon's Secret Service

It's nice to have error correction that is mathematically provable, but it's not strictly essential. As the first users of turbocodes found, if it works, that's good enough. But there is an area of information theory where mathematical proof is everything: cryptography.

Cryptography, the practice of creating and decoding secret communications, is perhaps the most under-appreciated branch of mathematics. Our freedoms and our prosperity are rooted in our ability to maintain privacy. Privacy is crucial, after all, for the operation of government and for online shopping. It enables secure mobile banking, which enables Rwandan farmers to conduct business and create a livelihood. It's central to Colombian anti-trafficking agencies co-ordinating a cocaine bust. Whistleblowers attempting to expose corruption need encrypted messaging services. Encryption is a vital resource — the oxygen of the information age.

Thanks to his wartime experience, Shannon quickly realised that the mathematics of information theory could shed light on just how well — or badly — an encryption system would work. In 1949, he published his ideas in 'Communication Theory of Secrecy Systems', which was a modified version of a classified document he had written in 1945.[28] The paper focuses on the situation where 'the message to be enciphered consists of a sequence of discrete symbols, each chosen from a finite set. These symbols may be letters in a language, words of a language, amplitude levels of a "quantised" speech or video signal, etc.' He pointed out that secrets encoded in symbols — unlike those hidden using invisible ink or encryptions requiring dedicated technologies such as a machine that can play voice recordings backward — can be analysed using mathematics. And, most importantly of all, he showed that mathematical analysis can tell you whether it's worth trying to bust them open. In other words, he worked out the mathematics of codebreaking, and of whether the effort is going to pay off.

This is hugely important, because it tells you where to direct your efforts. Follow Shannon's instructions properly, and you can change history — as the Special Fish Report makes clear.

Though it was sent to the US War Department in December 1944, the title of this report doesn't make it sound like something that should be marked 'top secret'. However, open its covers and it quickly reveals itself as an update on the state of efforts to break 'Fish', the name that British crytographers gave to the encrypted messages sent by German radio operators during the Second World War.[29] The author of the report was Albert Small of the US Army Signal Corps, who had been seconded to help with efforts at the UK's Bletchley Park codebreaking centre. Clearly, he was impressed. In the first paragraph of his update he says there was success on a daily basis. He put it down to 'British mathematical genius, superb engineering ability, and solid common sense', and called it 'an outstanding contribution to cryptanalytic science'.

But was it outstanding enough? The primary goal was to crack the Germans' Lorenz cipher, the even more fiendish successor to Enigma. Lorenz was theoretically able to generate perfectly random cryptographic keys. These were mixed with 'plaintext' typed messages by using developments of George Boole's logic and Shannon's expansion of it in his master's thesis: a valve-driven combination of AND, NOT, and OR gates that, together, made an XOR gate.

In theory, the result would be an unbreakable cipher. The Allies' only hope was that the implementation of the cipher was less perfect than the theory. And it was. The way the German telegraph operators used Lorenz had various shortcomings, and the way the machines themselves were set up created other chinks in its armour.

Enter the Colossus. Developed by a telephone engineer called Tommy Flowers, Colossus was the world's first programmable electronic digital computer, the ultimate ancestor of the machines with which we are so familiar.[30] It used XOR gates too, and was capable of performing 100 billion Boolean operations without making an error. Its inputs were fed by teleprinter tape running at nearly 30 miles per hour. All of this ingenious engineering meant that when it entered service on 5 February 1944, breaking the ever-changing Lorenz cipher began to take hours rather than days. But Flowers knew he could do even better, and on 1 June, Colossus Mk II took over. It was so fast that its reworked innards matched the operational speed of the first microchip that would be introduced by Intel three decades later.

Colossus Mk II was central to the success of the D-Day landings on the beaches of Normandy. It was used to decrypt radio messages between Hitler and his generals. Four days after its commissioning, a Bletchley Park courier brought General Dwight D. Eisenhower a note during a meeting with his staff. The note said that the Allies' various military deceptions surrounding D-Day had worked. According to intelligence reports uncovered using Colossus, Hitler believed the inevitable assault would happen to the east and had diverted huge numbers of troops to those regions, far from where the landings were due to take place. With the US 1st Army due to land on the westernmost of the chosen beaches, the report must have made Eisenhower a very happy man. He turned to his staff, and announced, 'We go tomorrow.' Many years later, Einsenhower declared that the codebreaking efforts at Bletchley Park had shortened the war by two years, saving hundreds of thousands of lives. George Boole, perhaps even Gottfried Leibniz, would no doubt be proud.

Perfecting Privacy

The story of mathematical codebreaking didn't actually start with Shannon. The best known origin is around AD 850, when the Arab mathematician and philosopher Abu Yusuf Ya'qub ibn Ishaq al-Kindi performed a statistical analysis of information in his 'Manuscript on Deciphering Cryptographic Messages'. You could often read the contents of an encrypted text, he showed, through statistical techniques such as frequency analysis. If you know which letters or words are most common (such as the letter 'e' in English words), you can find their substituted equivalents in the encrypted message and begin to break the code.

Since al-Kindi, people trying to keep secrets have always had to find new, unanticipated ways to make those substitutions. Ultimately, though, there's only one way to do it with confidence: develop a code where the substitution algorithm for encrypting and decrypting the message — the 'key' — can never be guessed. Such a perfect key will be entirely random in its substitutions, have at least as many characters or bits as the message, be known only to the sender and receiver, and be used only once so that there is no opportunity to perform statistical analysis. It's known in cryptography circles as the 'one-time pad'.

In his paper, Shannon was able to show that the only provably secure encryption methods are all mathematically equivalent to the one-time pad. But, although it is unbreakable, the one-time pad is terribly inconvenient. How do you ensure that the sender and receiver are the only ones who have access to this perfect cryptographic key? Either you need truly trustworthy couriers, or for sender and receiver to meet to share a key before going their separate ways. Unless they meet each time they want to communicate (in which case they could just whisper in each other's ear), they will need to share a whole set of keys that are to be stored and used only when necessary. Then they have to ensure that the storage method is entirely secure, and that they know which key will apply to which message.

These practical issues mean that the only mathematically secure method of encryption is rarely used. Instead, everyone employs imperfect codes. It's not such a terrible idea, given that a bigger problem tends to be the imperfect people that use them. As with the Lorenz cipher, the main reason Polish mathematicians had been able to break the German Enigma code (a breakthrough they conveyed to British intelligence operatives) was not because the Enigma machines were less than perfect, but because human operators fell into repetitive or guessable routines — many messages contained the sign-off 'Heil Hitler', for example.

So the intriguing question is this: how much do the shortcuts necessary for practicable one-time pad encryption compromise its security? The issues involved include the variety of 'choices' of characters available, the size of the key used to perform the encryption, and the number of encrypted messages that are intercepted. Shannon imagined trying to break a code with a brute-force attack that tries every possible combination of random keys while looking at the output for words and phrases that make sense. Then he defined the 'unicity distance' as the number of encrypted characters you would need to intercept in order to achieve this. This distance depends on what choices can be made when selecting a key, and on the statistical characteristics of the language. If a message in English is sent using a simple substitution code, he calculated that you need about 30 characters to be able to decipher it.

Thirty characters is not much, is it? That's why no one these days works with the simple codes that formed the basis of Shannon's examples. So what do they do instead?

The answer may surprise you. Although modern cryptography is fiercely complex, the state of the art is based on a surprisingly simple premise that takes us all the way back to Chapter 1. It's this: multiplication is easier than division.

If I ask you to tell me the result of multiplying 3 by 7, you'll almost instantly reply with 21. If I had asked you for the factors of 21, on the other hand — the integer numbers that multiply together to make 21, you would have had to think a little bit harder.

What if I were to ask you for the factors of 302,041? For this, the only thing you could do is use a brute-force approach, where you go through the options. You might start at 3 times one hundred thousand and something until you find the right combination. I say combination, not combinations, because there is only one answer to this example (besides itself and 1, that is): 302,041 is the result of multiplying 367 by 823. These two factors can't be divided down any further, because they are two of the infinite number of prime numbers, which are only divisible by themselves and 1. As with π and e, people get very mystical about prime numbers, loading them with all kinds of metaphysical baggage. What sometimes gets lost in the hocus-pocus is that they are of enormous practical use, too — especially if you are in the business of keeping secrets.

Encryption with primes started at Bell Labs — where else? In October 1944, an engineer called Walter Koenig Jr finished writing a classified document known as 'Final Report on Project C-43'.[31] The project, which had run in parallel with the X-System project that Shannon worked on, was a three-year investigation into speech-scrambling technologies.

'The immediate pressure behind these studies was caused, of course, by the War,' Koenig says in the introduction. The Army, Navy, and NDRC (National Defense Research Committee) wanted to know how they could best keep their voice communications secure, and how much of the enemy communications might be decrypted. Koenig is aware that, although this is the final report, there is much more to be done here. He recommends that, 'to keep up with the ever-changing art of communication, these studies should be continued under government auspices during peacetime.'

That wish came true. In 1969, an engineer called James Ellis stumbled across the report during his own research. Ellis was working for the British Government Communications Head Quarters (GCHQ), looking into ways to make encryption technology more practical. He found that Project C-43 had examined, among other things, the security offered when only one party injects noise into a telephone call. If the receiver sends an overwhelming amount of random electrical noise down the telephone line, and makes separate recordings of the call and the noise they injected, they can subtract out the noise later. An eavesdropper won't know the form of the noise, and so won't be able to separate the noise from the voice they are interested in. It's a 'one-way' function: easy to create, but impossible to reverse — unless you have the key.

Ellis was intrigued by the possibility of security created by just one party to the secret, and thought there must be a way to create a similar technology for transferring data. One summer's evening, he went to sleep and, as he said later, 'It was done in my head overnight'.[32] Being a good spy, he didn't write it down at home; he just hoped he would remember it.

And he did. In July 1969, Ellis's report hit the desk of GCHQ's chief mathematician, Shaun Wylie. Wylie's response gives an insight into the glass-half-empty mindset of an intelligence chief: 'Unfortunately,' he said, 'I can't see anything wrong with this.'

Perhaps to Wylie's relief, Ellis's idea was almost impossible to implement using the technology of the time. It wasn't until 1973, when a Cambridge mathematician called Clifford Cocks joined GCHQ, that a way forward came into view. Cocks had been doing postgraduate research on large prime numbers. When someone explained to him the basics of Ellis's idea, it immediately occurred to him that it would be possible to use prime numbers to reproduce the 'one-way' effect of adding noise to the telephone line.

He worked it all out in one evening. Because he was at home, he didn't write anything down, but his scheme stayed neatly in his head too. A (very) simplified version is this: Cocks performs a mathematical operation that includes two large primes to generate his 'public key'. He can make this publicly available so that someone wanting to send him a secret can mathematically blend their secret with the public key. They then send the resulting string of data to Cocks. Because Cocks is the only person who knows the maths that used two prime numbers to create the public key, he alone can decrypt the message and reveal the secret.

Ellis and Cocks wrote up their conception of 'public key cryptography' — but only for the eyes of British and American security services. A few years later, academic mathematicians also discovered the idea, which eventually became a commercial product: the 1977 Rivest–Shamir–Adleman (RSA) cryptosystem. Another 20 years passed before GCHQ revealed that they had actually discovered public key cryptography decades earlier.

Since Ellis and Cocks, creative mathematicians have devised a whole host of new ways to protect secrets. Reliable cryptography is so easy to implement now that these schemes protect our personal data, our credit card details, our communications, and anything else we choose to keep private. Online shopping tends to use public key cryptography, but Apple uses an encryption algorithm based on the mathematics of an 'elliptic curve' to lock its mobile devices. Instead of working from prime numbers, elliptic curve encryption uses the points on a graph to hide data. The algorithm defines a series of simple operations that will move you between different points on the curve; the eavesdropper only knows the end point and start point, and can't find the in-between points that hide the data. WhatsApp has a different approach: an algorithm called Signal Protocol, which is a combination of several encryption techniques. The only problem is, all of them are under threat because of a revolutionary, quantum version of codebreaking.

Information and the Quantum Future

We touched on the 'quantum' world of molecules, atoms, and subatomic particles when we looked into the strange worlds opened up by imaginary numbers. The rules by which they operate are rather different from the rules of everyday stuff. When information theory is implemented on a standard, or 'classical', computer, the binary digits are well-defined 0s and 1s. If you decide to encode your bits in a quantum computer, however, things can get a little fuzzy. And this, it turns out, makes all the difference.

In classical computers, the 1s and 0s are encoded as the particular state of an electrical circuit. It might be a voltage (or absence of a voltage), or the on/off state of a transistor, or the charged/uncharged state of a capacitor. In quantum computers, it's not so concrete. Here we encode

0s and 1s in entities that we can only describe through mathematics. As we discovered when we looked at imaginary numbers two chapters ago, the mathematics of the quantum world uses complex numbers and wave equations, and its physical manifestations aren't strictly of this world. And this means strange things can happen to the information.

In 1994, a mathematician working at a spin-out of — you guessed it — Bell Labs showed just how strange this can get. Peter Shor was looking at the mathematics of factoring: finding the two numbers that multiply together to give a larger, known number. As we have seen, traditional mathematics knows of no quick route to factoring: you just do it by trial and error. Quantum mathematics, however, has a trick up its sleeve.

It's complicated, but it's best explained by talking about the quantum entities encoding the information as waves. These waves, like ripples on a pond, can 'interfere' with each other: where the ripples meet, their structures become altered in predictable ways. Waves have another property too: some of their attributes, such as location, don't have a precise, well-constrained answer. Shor showed that the unknown factors can be found by manipulating the interference between the undefined properties of the wave. A fuller explanation involves Fourier transforms; the essential point is that a sufficiently large quantum computer, one that encodes a lot of quantum bits (qubits) simultaneously, can use Shor's algorithm to find the prime factors of large numbers with consummate ease.

This discovery created significant ripples inside national security bodies across the globe. In the years that followed, governments poured vast sums of money into the study of quantum computing. They needed to find out how easy it was going to be to build these machines, and whether they really were going to be as problematic as Shor's algorithm implied. The truth is, progress in building quantum computers has been slow; it was only two decades later, in 2016, that the US National Security Agency issued a statement on the issue. 'NSA does not know if or when a quantum computer of sufficient size to exploit public key cryptography will exist,' it said. However, the statement continued with a note of caution: 'There is growing research in the area of quantum computing, and enough progress is being made that the NSA must act now.' The agency advised all US businesses to move away from encryption that was based on the factoring of large numbers. It was clear that the RSA, elliptic curve, and other systems might soon be as useful as a bucket made of holes.[33]

Perhaps you'll be reassured to know that Shannon's encryption work is still being extended: some of today's best mathematicians are now developing new, replacement algorithms that will withstand even a quantum attack. And other mathematicians have reconfigured Shannon's 1949 work on cryptography so that it is fit for this new quantum information age. Going back to the incomparable, uncrackable one-time pad, they have harnessed the powers of the quantum world to give us a new way to distribute those encryption keys safely. Quantum cryptography, as the result is known, is a means of sending the bits for cryptographic keys down an optical fibre, or around the world by satellite, with perfect security. If an eavesdropper intercepts them, or even tries to skim off just a fraction of the key, the mathematical laws of the quantum world dictate that the sender and receiver will be able to tell. All they then have to do is repeat the key distribution process with a fresh set of numbers.

It's worth mentioning as a coda to this devastatingly practical, down-to-earth research that it has had an unexpected spin-off. Combining binary logic with the laws of quantum physics has stimulated a new, quantum-centred search for explanations of the universe, of human thought and patterns of behaviour. It's almost as if we are developing a quantum version of the I-Ching; Leibniz would be delighted.

At the heart of this research effort is the curious phrase 'It from Bit'. It was coined by physicist John Wheeler, the same man who gave us the term 'black hole'. Wheeler's 'It' is everything around us: the cosmos. The

'Bit' is Shannon's binary digit. Wheeler expressed his ideas in a formal research paper entitled 'Information, Physics, Quantum: The Search for Links', and its first line would have thrilled Leibniz and Boole: 'This report reviews what quantum physics and information theory have to tell us about the age-old question, How come existence?'[34]

Wheeler explained It from Bit as the 'most effective' formulation of the idea that 'every it — every particle, every field of force, even the spacetime continuum itself — derives its function, its meaning, its very existence entirely — even if in some contexts indirectly — from the apparatus-elicited answers to yes or no questions, binary choices, bits.' It seemed reasonable to Wheeler to reduce everything in the universe to information that comes to us in the form of a binary digit. Put quantum theory and these simple building blocks of information together in the right way, and you get space and time, stars and planets, and you and me.

The quest continues: these days, the physicists trying to understand the complexity of the universe suggest that information theory is the landscape to explore. They are thinking in terms of information 'entropy', which means mapping and quantifying the transmission of information, and of computing, where every law of physics and chemistry can be recast as a computation that processes the bits of the physical universe in quantum versions of logic gates. We are a result of those computations, and our thoughts and actions contribute to their unfolding. As the physicist Seth Lloyd has put it, 'every atom, every elementary particle participates in the huge computation that is the universe', and 'every human being on Earth is part of a shared computation'.[35] At the cutting edge of physics, everything in the universe — including us — can be reduced to a processing of Shannon, Boole, and Leibniz's bits: true and false; yes and no; 1 and 0.

The Greatest Showman

I want to finish this chapter by bringing the man at its centre a little more into focus. In previous chapters, we have seen some less than enticing characters — I'm thinking of Newton and Descartes in particular — and it would be a shame to leave you with the impression that mathematical geniuses are always disagreeable.

It's hard to find a bad word said about Claude Shannon. Like anyone with a head full of thoughts, he could sometimes be difficult to engage, but he was also unremittingly playful. As a child, Shannon had fantasised about being a fairground performer, and became obsessed by the hard-learned motor skills of juggling and just how difficult they were to perfect. That's why he learned to juggle for himself — and went on to design and build robots that could juggle. His machine-clowns were so precisely engineered that Shannon would boast that they 'juggle all night and never drop a prop!'[36]

His own fallibility pushed him hard to new heights: he learned to ride a unicycle, and then to juggle while on the unicycle. Then he learned to juggle while unicycling on a steel tightrope. And this is only scratching the surface of his activities away from information theory. He also made polystyrene shoes for walking on water and an enormous beckoning finger that would summon his wife out of the kitchen when he flicked a switch in his basement lab.

The beckoning finger was a joke machine — Betty might have been a great cook, but she was also a superb scientist whose collaboration was invaluable to Shannon's work. There were other joke machines too: Shannon built a flame-throwing trumpet, for instance, and the first 'Useless Machine'. If you've never seen this device, it's wonderful: its only purpose is to turn itself off. You flick the switch to 'on', and an arm comes out of a closed box to flick it back to 'off'. The arm retracts, and the machine powers down until you switch it on again.

Shannon got the idea from the computing and robotics pioneer Marvin Minsky. It is a great testimony to Shannon's personality that, once he had heard the idea, he couldn't resist making it a reality. Not everyone was amused by the Useless Machine, though. The science fiction writer Arthur C. Clarke found it disturbing. 'There is something unspeakably sinister about a machine that does nothing — absolutely nothing — except switch itself off,' he said.[37]

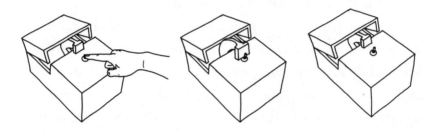

Claude Shannon's Useless Machine, which does nothing but turn itself off

Another of Shannon's machines had much more *raison d'être*. It was the first wearable computer, and it was designed to analyse the speed and trajectory of a ball on a roulette wheel.[38] The size of a cigarette packet, it was connected to a series of microswitches in the wearer's shoes. This was how the operator would reset the machine, and start the process of analysis. The main input was a foot-switch signal giving the time it took the ball to complete one revolution once the game was in motion. The output was a musical tone, played into a wire-fed earpiece, that told the wearer how to bet.

In the summer of 1961, Shannon and the machine's co-developer, his graduate student Edward Thorp, took their wives to Las Vegas to try it out. The wives were the lookouts, checking on whether anyone was getting suspicious. Generally, they weren't — except on the occasion that an earpiece's feed wire broke, leaving Thorp's earpiece peering out from his ear canal 'like an alien insect', as he later recalled. Apart from a few broken wires, the computer performed well, though. Thorp and Shannon even considered a return trip with their hair grown longer to better conceal the earpiece.

They never went back. In fact, Shannon began to disappear from his colleagues' view in the 1960s. By the end of the decade, he had stopped attending conferences on information theory. But he had not disconnected from the world, or from his friends. He made a series of investments in companies started up by people he knew and trusted. Former colleague Bill Harrison was one, and when Harrison Laboratories were bought up by Bill Hewlett and David Packard, Shannon found himself an early stockholder of HP. Then there was his MIT classmate Henry Singleton's company, Teledyne. Shannon invested because he respected Singleton's abilities, and his instinct paid off: Teledyne went on to become a multi-billion-dollar company. The fact that he ended up with early Motorola stock came from the same instinct to help his friends get their ideas off the ground.[39]

Despite disappearing from public view, Shannon's popularity never waned. That was clear when he made a surprise appearance at a conference in Brighton, England. It was 1985, and Shannon was 69 years old. For some reason that no one now remembers, Shannon was wandering in and out of talks at Brighton's Grand Hotel during the International Symposium on Information Theory. Someone recognised him, and whispers began to circulate that the father of information theory was actually there. The conference organiser, Robert J. McEliece, later summed up the atmosphere: 'It was as if Newton had showed up at a physics conference.'[40]

It's a nice analogy, but not many of Newton's peers would have wanted to spend time with him. Shannon, though, was liked as well as admired. He was quickly trapped, and the conference organisers pressed him into making a speech at the conference dinner. When the time came, Shannon worried he would bore everyone. So he said a few words, pulled out some juggling balls and turned his speech into a cabaret performance. The night ended with physicists, who generally have no interest in celebrities, joining a long queue to get Shannon's autograph.

Claude Shannon died in 2001. Ironically, it was Alzheimer's disease that finally tore down this Colossus: a lifetime of carefully stored and catalogued information was gradually erased from his plaque-ravaged brain. It was a sad end for an extraordinary life that had been crammed with achievement.

Shannon's information theory was such a profound and impactful idea that it changed the human experience almost immediately. In fact, in 1956, eight years after information theory first saw the light, Shannon felt compelled to discourage people from applying his work too widely and wildly. 'Applications are being made to biology, psychology, linguistics, fundamental physics, economics, the theory of organization, and many others,' he wrote, rather disapprovingly, in an essay entitled *The Bandwagon*.[41] Although he admitted this 'heady draught of general popularity' was rather 'pleasant and exciting', he insisted that information theory would not be applicable to everything. 'Seldom do more than a few of nature's secrets give way at one time,' Shannon said.

It's an extraordinary paper. How often does a mathematician need to tell people to back off and stop trying to apply his work in the real world? But you can hardly blame them: there seems to be no area of life that can't benefit from Shannon's insight. It has taught us the secrets of the solar system and how to enjoy worry-free online shopping. It brings movies on demand and (hopefully) the final theory of physics. It ties together the internet and the I-Ching. Whether we consider war-winning computers, data-loaded mobile phone signals or songs streamed and beamed through the air and into our ears, our world would be unrecognisable without Shannon's contribution to mathematics. Information theory is the culmination of tens of thousands of years of human insight, invention, and ingenuity; the pinnacle of the art of more.

CONCLUSION

Maths is a many-splendored thing

You and I would probably agree that we are both civilised. But what does that mean? Scholars rarely agree on what exactly defines a civilisation, but they do tend to offer some commonly accepted characteristics. A civilisation will have large settlements — cities, effectively. Its society will incorporate some form of religion. There will be division of labour, specialisation of skills and a form of central government by established law, almost certainly with some kind of taxation to pay for its administration. There will be a class system and a stable food supply. Some citizens of a civilisation will have leisure time, which gives space for art, music, and other culture to develop.

Most studies claim that a culture of writing is also an essential component of civilisation. However, we know that the Incan Empire — surely one of the great civilisations — didn't have any form of written language. But there is something that the Inca *did* have that always seems to get left off everyone's list of what makes a civilisation. This should, in fact, be the first — perhaps the only — requirement. I'm talking about mathematics, of course.

The Inca recorded government data, trading records, accounts, and many other sets of numbers in knotted strings called *quipus*. Every town had a 'keeper of the knots' who was appointed by the king and acted as a government statistician — much like the samurai in Japan. We have seen mathematics at work at the heart of the Sumerian kingdoms five thousand years ago, and in the development of African civilisations, both in the north and in sub-Saharan Africa. In the early 14th century, Mansa Musa, reportedly the richest man who ever lived, built a vast university in Timbuktu that taught mathematics alongside astronomy and law. Musa's Malian Empire, the source of most of the gold in circulation in the medieval world, was built on trade and taxation and owed everything to a mastery of numbers.

And we still owe that same debt, seven centuries on. Let's make a short list of the good things mathematics has brought us: global travel; supermarket shelves bursting with produce; refrigeration; mobile phones; complex and beautiful urban environments; the entertainment industry; access to finance that has engendered unprecedented prosperity; phenomenal works of art; many extra decades of healthy life; profound knowledge of the cosmos and its history; the extraordinary resource that is the internet — and that's just off the top of my head. All of which should make us wonder why the profound influence of maths has remained hidden for so long.

I blame Plato. Our world, this Greek philosopher declared in the 4th century BC, is the shadow of a perfect reality composed of mathematical ideals. He espoused the idea that the universe was built upon a framework defined by a handful of solid geometrical shapes. Chief among them was the 12-sided dodecahedron; he described it as the shape that God used 'as a model for the twelvefold division of the Zodiac'.[1]

Around 300 BC, Euclid used Plato's worldview when he wrote *Elements*, his summary of mathematics. *Elements* has been described as the most influential textbook ever produced, but it included no attributions, no discussion of where the ideas came from, and how humans had developed them. It was as if mathematics had been handed down to us on stone tablets. As a result, mathematics was taught for century after century as a subject almost akin to theology. You only have to examine the fuss surrounding the 'golden ratio' to see this.

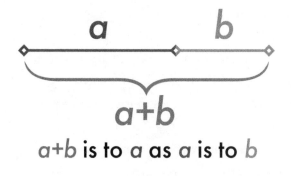

A line divided in the golden ratio

The golden ratio is rather simple to describe: divide a line in two so that the ratio of the whole line to the larger part is the same as the ratio of the larger part to the smaller one. The ratio is$(1 + \sqrt{5})/2$; roughly 1.618. Such was the mysterious power attributed to this number that, when Luca Pacioli published a book about it in 1509, he called it *The Divine Proportion*. The contents page is enough to convey Pacioli's awe: Chapter 5, for instance, concerns 'the fitting title of the present tract'. That is, it will explain why the proportion under discussion is indeed divine. Chapters 11–14 are about the proportion's properties, which are (in order) 'essential', 'singular', 'ineffable' and 'marvellous'. By Chapter 15 the described property is 'unnameable'; it then becomes 'inestimable'. Then there is a 'supreme', a 'very excellent' and an 'almost incomprehensible' property. An author's enthusiasm for their subject has perhaps never seemed so clear.

As it began, so it continued. It only became known as the 'golden ratio' in the 19th century, but because Pacioli's friend (and student) Leonardo da Vinci created the illustrations for his book, scholars have projected the golden ratio onto the features many of da Vinci's works, including the *Mona Lisa* and *Vetruvian Man*. Some have claimed, for instance, that the proportions of the Mona Lisa's face follow the proportions of the golden ratio. None of these claims stand much scrutiny; they all depend on how you take the measurements.[2]

Attempts to find the golden ratio in architecture are similarly problematic. People have claimed that the great pyramids, various cathedrals, and the Parthenon have all been designed to conform to the strictures of the golden ratio, but most researchers are more than sceptical. Nonetheless, its mythical power overcame modern architects such as Le Corbusier, who felt it, and the Fibonacci series, to be so primal that they could be considered as 'rhythms apparent to the eye'. They are, he said, at the very root of human activities: 'They resound in man by an organic inevitability, the same fine inevitability which causes the tracing out of the Golden Section by children, old men, savages and the learned.'[3]

There is little truth to Le Corbusier's pompous musings. The golden ratio does exist in nature: it determines the properties of a vast number of natural phenomena, from the way leaves are arranged on a plant stem to the proportions of a spiral seashell and the thermodynamic properties of black holes. But there is no need to get mystical about it; many other numbers occur repeatedly in natural phenomena.

One place where the golden ratio's existence is undeniable is in Salvador Dalí's *The Sacrament of the Last Supper*. The picture is painted within a rectangle whose sides are proportioned in the golden ratio. And behind Christ and his apostles sits an enormous dodecahedron; a geometric shape whose properties also reflect the golden ratio. This, however, is entirely deliberate. Dalí's choice is not based on aesthetics, but symbolism: he is kneeling before Plato. It's the kind of genuflection that caused others to ascribe quasi-mystical powers to various mathematical components. I have already mentioned that I don't find π, e or the square root of 2 particularly inspirational. I feel the same about prime numbers, too.

Let me put it like this. As we have seen, humans invented the integers — the whole numbers — to describe and manipulate things we found (or imagined) in our environment. We also invented the concept of division to enable us to share those things out. Unsurprisingly, we found that some of the whole numbers can be divided by other whole numbers to give yet other, smaller whole numbers. At the same time, some of them — the prime numbers — cannot. Why should this surprise us? It's just a consequence of the way numbers work. It's interesting that these prime numbers arise in particular places on the number line that we imagine stretching from 1 to infinity. But it's not a mystery. Calling it a mystery would be like ascribing supernatural power to the saffron used to create a particularly tasty meal. Yes, you can describe a spice as an expensive, fragrant, ancient ingredient brought to us centuries ago from the mysterious East. But you can also describe it as a carrier for the chemicals crocin and picocrocin, which interact with other chemicals in food to create a golden yellow meal with a particularly pleasant taste that people have enjoyed for thousands of years.

Perhaps this feels like a sacrilege — as though I am attempting to 'unweave a rainbow'. This was the accusation raised by John Keats against Newton's demonstration that the various colours of light combine to produce white light. But I think there is good reason to unweave the mathematical rainbow. As things stand, those who have been initiated into its Platonic cult are the only ones who see the power of numbers. But if we demystify mathematics, perhaps we can democratise it. Everyone might finally have the chance to see that maths is composed from useful threads, none of which require a special kind of mind to appreciate their value. They may even begin to appreciate that grasping and working with some of these threads can be enjoyable and fruitful. Mathematics should be for all, surely?

It's hard to know whether the gradual appropriation of mathematics by an elite minority was deliberate. The ancient Egyptians' nilometers suggest it was: these gauges of the river's depth were built within the boundaries of the temples to give the priests exclusive access. They alone knew when the floods were coming, and so possessed secrets that affected the lives of ordinary people, an important attribute when seeking power over the masses. But even if the mathematicians of history weren't explicitly looking to gain power, it's not difficult to imagine an unconscious desire to present their subject as something deep and powerful, and difficult to access. Translated into the terms of the mathematically derived discipline of economics it makes perfect sense: it's a simple case of creating demand for something that you alone can supply.

There is an alternative to elitist, mystical mathematical thinking. Instead of seeing mathematicians as explorers making discoveries in the Platonic landscape, perhaps we should see mathematicians as artists who create the subject. They are painting from a palette that overflows with numbers, their tool tray balancing an ever-growing array of algorithmic knives and brushes. Most are completing works begun long ago, filling in gaps left by the old masters. But sometimes a mathematician will paint something entirely new and startling. That is how we ended up with works like geometry, the logarithm, information theory, and a solution to Fermat's last theorem. The real beauty of it is that these creations, unlike a genius's painting, belong to all of us. With the new maths, we create stunning architecture, a life-saving medical treatment, a data compression technology that brings joy to millions, a scientific breakthrough that shows us our place in the universe — or any number of the other achievements of human history.

Our story is inextricably entwined with the story of mathematics. We counted, and invented money and trade. We drew shapes in the sand, and learned how to travel safely around the globe. We used what we knew to figure out what we didn't, and built a complex, networked, interdependent society that allowed some people to spend their time filling in the gaps, creating new opportunities to create wealth and prosperity. We saw how the properties of triangles and circles would bring hitherto impossible calculations to heel, and used the resulting tools to engineer our way into the twentieth century. We understood that abstract ideas like information and imaginary numbers were the key to unlocking atomic, electrical, and electronic power; you are living through the marvel-filled consequences of that. Mathematics has shaped the very experience of what it means to be human, and left its mark on all of us — we just didn't see it until now. So, while we may never agree whether we discovered mathematics or created it, perhaps we can now all agree on something: that mathematics created us.

Acknowledgements

Finishing a book, and finally presenting it to the world, is bittersweet. I don't think I have ever enjoyed writing a book as much as this one, and I feel a hint of sadness that the fun of learning so many new things about numbers is finally over.

My family are probably relieved, however. No more evenings of me exuberantly assailing them with the fruits of that day's research; no more 'try this Egyptian method of doing multiplication for me — it won't take you a minute ...'; no more 'tell me what it was like to encounter imaginary numbers for the first time ...'; no more 'I think I want a slide rule/sextant/abacus for my next birthday ...' Phillippa, Millie, Nova — thank you for your patience. I hope it wasn't too painful.

The fact that this book exists at all is down to Patrick Walsh and the team at PEW Literary. I was astonished when Patrick jumped at my vague, half-baked idea of a book on maths for people who had never quite managed to engage with the subject. We fleshed it out together while walking on the clifftops above Cuckmere Haven, and I have never doubted the idea since. My editors Molly Slight and Edward Kastenmeier, and their teams at Scribe and Pantheon respectively, have provided insightful and enthusiastic support throughout this project. Special thanks to copyeditor Richard Leigh for his painstaking work on the manuscript and for enjoying the Marshall Plan joke, and to Philip Gwyn Jones for his early support.

I am extremely grateful to my honest and diligent readers: experts Artur Ekert and Matthew Hankins, and non-experts Shaun Garner and Charlie Higson. And somewhere in between — he knows what I mean — was Rick Edwards. All of them provided valuable help, but any errors that remain (of fact or judgement) are mine.

Finally, I want to thank those who helped me out with insights on a range of subjects, from the history of accounting to the day-to-day practice of architects. They are, in no particular order, Richard Lindley, Jens Hoyrup, Jon Butterworth, Melanie Bayley, Christopher Napier, Keith Hoskin, Richard Macve, Manfred Zollner, Jean-Jacques Crappier, Erdal Arikan, Radford Neal, Nick Kingsbury, David Blockley, and Andrew Whitehurst.

Michael Brooks
May 2021

Notes

Introduction

1. Peter Gordon, 'Numerical cognition without words: evidence from Amazonia', *Science* 306, no. 5695 (15 October 2004): 496–99, https://doi.org/10.1126/science.1094492.
2. Caleb Everett, *Numbers and the Making of Us: counting and the course of human cultures* (Cambridge, MA: Harvard University Press, 2019).
3. Rachel Nuwer, 'Babies are born with some math skills', *Science | AAAS*, 21 October 2013, https://www.sciencemag.org/news/2013/10/babies-are-born-some-math-skills.
4. John Dee, *The Mathematicall Praeface to Elements of Geometrie of Euclid of Megara*, http://www.gutenberg.org/files/22062/22062-h/22062-h.htm.

Chapter 1: Arithmetic

1. Richard Brooks, *Bean Counters: the triumph of the accountants and how they broke capitalism* (London: Atlantic Books, 2019).
2. François-Auguste-Marie-Alexis Mignet, *History of the French Revolution from 1789 to 1814*, https://www.gutenberg.org/files/9602/9602-8.txt.
3. Jacob Soll, *The Reckoning: financial accountability and the rise and fall of nations* (New York: Basic Books, 2014).
4. Founders Online, 'From Alexander Hamilton to Robert Morris, [30 April 1781]', http://founders.archives.gov/documents/Hamilton/01-02-02-1167.
5. There is another bone that might have a claim to being an older mathematical artefact. Known as the Lebombo Bone, it is thought to be

approximately 43,000 years old, and features some notches that might be tally marks. There is considerable doubt over this, though, and its discoverer, South African archaeologist Peter Beaumont, certainly does not make the claim that this is a mathematical tally.

6. Thorsten Fehr, Chris Code, and Manfred Herrmann, 'Common brain regions underlying different arithmetic operations as revealed by conjunct fMRI-BOLD activation', *Brain Research* 1172 (3 October 2007): 93–102, https://doi.org/10.1016/j.brainres.2007.07.043.

7. Simone Pika, Elena Nicoladis, and Paula Marentette, 'How to order a beer: cultural differences in the use of conventional gestures for numbers', *Journal of Cross-Cultural Psychology* 40, no. 1 (1 January 2009): 70–80, https://doi.org/10.1177/0022022108326197.

8. Georges Ifrah, *From One to Zero: a universal history of numbers* (New York: Penguin Books, 1987).

9. Ilaria Berteletti and James R. Booth, 'Perceiving fingers in single-digit arithmetic problems', *Frontiers in Psychology* 6 (16 March 2015), https://doi.org/10.3389/fpsyg.2015.00226

10. Brian Butterworth, *The Mathematical Brain* (London: Macmillan, 1999).

11. Jens Høyrup, 'State, "justice", scribal culture and mathematics in ancient Mesopotamia: Sarton Chair Lecture', *Sartoniana* 22 (2009): 13–45.

12. Jens Høyrup, 'On a collection of geometrical riddles and their role in the shaping of four to six "algebras"', *Science in Context* 14, no. 1–2 (June 2001): 85–131, https://doi.org/10.1017/S0269889701000047. (The answer is 4.874. It can be worked out through the quadratic formula, which we haven't yet encountered.)

13. Crappier J.-J., Farinetto C., Gascou P., Maunoury C., Maunoury F. & Mateusen G., 'The Akan Weighing System restored after 120 years of oblivion. A metrological study of 9301 geometric gold-weights', *Colligo*, 2(2) (21 November 2019), https://perma.cc/H494-E42R.

14. E.W. Scripture, 'Arithmetical prodigies', *American Journal of Psychology* 4, no. 1 (1891): 1–59, https://doi.org/10.2307/1411838.

15. Sylvie Duvernoy, 'Leonardo and theoretical mathematics', in *Nexus Network Journal: Leonardo da Vinci: Architecture and Mathematics*, ed. Sylvie Duvernoy (Basel: Birkhäuser, 2008), 39–49, https://doi.org/10.1007/978-3-7643-8728-0_5.

16. If you feel sympathy with Leonardo, that's understandable. Of course, you can just accept that dividing a quantity by numbers less than 1 makes that quantity bigger. However, it might help to work through an example. Imagine dividing 10 chocolate bars between 5 ice hockey teams. Every team gets 2 bars. Now imagine dividing the bars between just 2 teams; each team gets 5 bars. As you decrease the number you're dividing by, the answer gets bigger. That continues as the number goes below 1. So let's go to numbers

lower than 1. Imagine dividing the 10 chocolate bars between just one-third of a team. One-third of an ice-hockey team is 2 people. So there are 10 bars divided between 2 people, giving each player 5 bars each. But that's equivalent to a full team having $5 \times 6 = 30$ bars of chocolate. So 10 divided by $\frac{1}{3}$ is 30.

17.   Julie McNamara and Meghan M. Shaughnessy, 'Student errors: what can they tell us about what students DO Understand?', Math Solutions, 2011, http://akrti2015.pbworks.com/f/StudentErrors_JM_MS_Article.pdf.

18.   The answer to the first question is $\frac{2}{3}$, $\frac{1}{2}$, $\frac{5}{9}$. The answer to the second question is 2. To reach this you can either approximate ($\frac{12}{13}$ and $\frac{7}{8}$ are both close to 1, so their sum is close to 2), or make the denominators the same and see what you get. We can turn $\frac{12}{13}$ into $\frac{96}{104}$ by multiplying top and bottom by 8. Then we turn $\frac{7}{8}$ into $\frac{91}{104}$ by multiplying top and bottom by 13. Then we add the two numerators. $96 + 91 = 187$, so the sum is $\frac{187}{104}$. That's approximately 1.8, so the closest option is 2.

19.   The Fibonacci series arises by starting with 0 and 1, and then adding the two previous numbers to get the next number in the series. The first 12 numbers in the series are 0, 1, 1, 2, 3, 5, 8, 13, 21, 34, 55 and 89.

20.   Blaise Pascal, *Pensées*, https://www.gutenberg.org/files/18269/18269-h/18269-h.htm.

21.   John Wallis, 'A Treatise of Algebra, Both Historical and Practical', *Philosophical Transactions of the Royal Society of London* 15, no. 173 (1 January 1685): 1095–1106, https://doi.org/10.1098/rstl.1685.0053.

22.   Charles Seife, *Zero: The Biography of a Dangerous Idea* (New York: Viking, 2000).

23.   Robert Kaplan, *The Nothing That Is: a natural history of zero* (Oxford: Oxford University Press, 2000).

24.   'The Internet Classics Archive | Physics by Aristotle', http://classics.mit.edu/Aristotle/physics.html.

25.   Jian Weng et al., 'The effects of long-term abacus training on topological properties of brain functional networks', *Scientific Reports* 7, no. 1 (18 August 2017): 8862, https://doi.org/10.1038/s41598-017-08955-2.

26.   Richard Goldthwaite, 'The practice and culture of accounting in Renaissance Florence', *Enterprise & Society* 16, no. 3 (September 2015): 611–47, https://doi.org/10.1017/eso.2015.17.

27.   Jane Gleeson-White, *Double Entry: how the merchants of Venice created modern finance* (New York: W.W. Norton & Co, 2012).

28.   Michael Schemmen, *The Rules of Double-Entry Bookkeeping (a Translation of Particularis de Computis et Scripturis)* (IICPA Publications, 1494).

29.   Steven Anzovin and Janet Podell, *Famous First Facts, International Edition: a record of first happenings, discoveries, and inventions in world history* (New York: H.W. Wilson, 2000).

30. Edward Peragallo, 'Jachomo Badoer, Renaissance man of commerce, and his ledger', *Accounting and Business Research* 10, sup1 (1 March 1980): 93–101, https://doi.org/10.1080/00014788.1979.9728774.

31. Allan Nevins, *John D Rockefeller: The Heroic Age Of American Enterprise* (New York: Charles Scribner's Sons, 1940), http://archive.org/details/in.ernet.dli.2015.58470.

32. Neil McKendrick, 'Josiah Wedgwood and cost accounting in the Industrial Revolution', *Economic History Review* 23, no. 1 (1970): 45–67, https://doi.org/10.2307/2594563.

33. Gleeson-White, *Double Entry*.

34. Ibid.

Chapter 2: Geometry

1. Andrew Kurt, 'The search for Prester John, a projected crusade and the eroding prestige of Ethiopian kings, *c.*1200–*c.*1540', *Journal of Medieval History* 39, no. 3 (1 September 2013): 297–320, https://doi.org/10.1080/03044181.2013.789978.

2. W.G.L. Randles, 'The alleged nautical school founded in the fifteenth century at Sagres by Prince Henry of Portugal, called the "Navigator"', *Imago Mundi* 45, no. 1 (1 January 1993): 20–28, https://doi.org/10.1080/03085699308592761.

3. Carl Huffman, 'Pythagoras', in *The Stanford Encyclopedia of Philosophy*, ed. Edward N. Zalta, Winter 2018 edition (Metaphysics Research Lab, Stanford University, 2018), https://plato.stanford.edu/archives/win2018/entries/pythagoras/.

4. Margaret E. Schotte, *Sailing School: navigating science and skill, 1550–1800* (Baltimore, MD: Johns Hopkins University Press, 2019).

5. E.G.R. Taylor, 'Mathematics and the navigator in the thirteenth century', *Journal of Navigation* 13, no. 1 (January 1960): 1–12, https://doi.org/10.1017/S0373463300037176.

6. James Alexander, 'Loxodromes: A rhumb way to go', *Mathematics Magazine* 77, no. 5 (2004): 349–56, https://www.tandfonline.com/doi/abs/10.1080/0025570X.2004.11953279.

7. 'The four voyages', in *Christopher Columbus and the Enterprise of the Indies: A Brief History with Documents*, ed. Geoffrey Symcox and Blair Sullivan (New York: Palgrave Macmillan US, 2005), 60–139, https://doi.org/10.1007/978-1-137-08059-2_3.

8. Mark Monmonier, 'The lives they lived: John P. Snyder; the Earth made flat', *The New York Times*, 4 January 1998, sec. Magazine, https://www.nytimes.com/1998/01/04/magazine/the-lives-they-lived-john-p-snyder-the-earth-made-flat.html.

9. John W. Hessler, *Projecting Time: John Parr Snyder and the development of*

the Space Oblique Mercator, Philip Lee Phillips Society Occasional Paper Series, No. 5 (Washington, DC: Geography and Map Division, Library of Congress, 2004), https://www.loc.gov/rr/geogmap/pdf/plp/occasional/OccPaper5.pdf.

10. Helge Svenshon, 'Heron of Alexandria and the dome of Hagia Sophia in Istanbul', Karl-Eugen Kurrer, Werner Lorenz, Volker Wetzk (eds), *Proceedings of the Third International Congress on Construction History (Cottbus 2009)*, Vol. 3, pp. 1387–1394, https://www.academia.edu/3177251/Heron_of_Alexandria_and_the_Dome_of_Hagia_Sophia_in_Istanbul.

11. Giulia Ceriani Sebregondi and Richard Schofield, 'First principles: Gabriele Stornaloco and Milan Cathedral', *Architectural History* 59 (2016): 63–122, https://doi.org/10.1017/arh.2016.3.

12. Krisztina Fehér et al., 'Pentagons in medieval sources and architecture', *Nexus Network Journal* 21, no. 3 (1 December 2019): 681–703, https://doi.org/10.1007/s00004-019-00450-7.

13. Samuel Y. Edgerton, *The Mirror, the Window, and the Telescope: how Renaissance linear perspective changed our vision of the universe* (Ithaca, NY: Cornell University Press, 2009).

14. Antonio Manetti, *The Life of Brunelleschi* (University Park, PA: Pennsylvania State University Press, 1970).

15. Marjorie Licht and Peter Tigler, 'Filarete's *Treatise on Architecture* (Yale Publications in the History of Art, 16), trans. with intro. by John R. Spencer', *The Art Bulletin* 49, no. 4 (1 December 1967): 351–60, https://doi.org/10.1080/00043079.1967.10788676.

16. Leon Battista Alberti, *On Painting* (London: Penguin, 1991).

17. Evelyn Lamb, 'The slowest way to draw a lute', Scientific American Blog Network, https://blogs.scientificamerican.com/roots-of-unity/the-slowest-way-to-draw-a-lute/.

18. Albrecht Dürer, *Memoirs of Journeys to Venice and the Low Countries*, trans. Rudolf Tombo (Auckland: Floating Press, 2010), http://search.ebscohost.com/login.aspx?direct=true&scope=site&db=nlebk&db=nlabk&AN=330759.

19. Kay E. Ramey, Reed Stevens, and David H. Uttal, 'In-FUSE-ing STEAM learning with spatial reasoning: distributed spatial sensemaking in school-based making activities', *Journal of Educational Psychology* 112, no. 3 (2020): 466–93, https://doi.org/10.1037/edu0000422.

20. Isabel S. Gordon and Sophie Sorkin, *The Armchair Science Reader* (New York: Simon and Schuster, 1959).

21. Michael Francis Atiyah, *Collected Works*, vol. 6 (Oxford: Clarendon Press, 1988).

Chapter 3: Algebra

1. 'FedEx History', FedEx, https://www.fedex.com/en-us/about/history.html.

2. Kent E. Morrison, 'The FedEx problem', *College Mathematics Journal* 41, no. 3 (May 2010): 222–32, https://doi.org/10.4169/074683410X488719.

3. John Hadley and David Singmaster, 'Problems to sharpen the young', *Mathematical Gazette* 76, no. 475 (1992): 102–26, https://doi.org/10.2307/3620384.

4. The one who asked for two oxen to be given him had four, and the one who was asked had eight. You can make 100 tunics from the large linen cloth.

5. Terry Moore, 'Why X marks the unknown', *Cosmos Magazine*, 14 June 2015, https://cosmosmagazine.com/mathematics/why-x-marks-unknown-0/.

6. Florian Cajori, *A History of Mathematical Notations, Volume I: Notations in Elementary Mathematics* (London: The Open Court Company, Publishers, 1928), http://archive.org/details/historyofmathema031756mbp.

7. Jens Høyrup, 'Algebra in cuneiform: Introduction to an Old Babylonian geometrical technique', Max-Planck-Institut für Wissenschaftsgeschichte, Preprint Vol. 452, 2013, https://forskning.ruc.dk/en/publications/algebra-in-cuneiform-introduction-to-an-old-babylonian-geometrica.

8. Will Woodward, 'Make maths optional — union leader', *The Guardian*, 22 April 2003, http://www.theguardian.com/uk/2003/apr/22/schools.politics.

9. House of Commons Hansard Debates for 26 Jun 2003, https://publications.parliament.uk/pa/cm200203/cmhansrd/vo030626/debtext/30626-22.htm, in col. 1264.

10. Ana Susac and Sven Braeutigam, 'A case for neuroscience in mathematics education', *Frontiers in Human Neuroscience* 8 (21 May 2014), https://doi.org/10.3389/fnhum.2014.00314.

11. Georg Christoph Lichtenberg, *Briefwechsel, Band III: 1785–1792*, eds Ulrich Joost and Albrecht Schöne (Munich: Beck, 1990).

12. Wilhelm Ostwald, 'Über Papierformate', *Mitteilungen des Normenausschusses der Deutschen Industrie* 12 (November 1918): 199–200 , https://www.cl.cam.ac.uk/~mgk25/volatile/DIN-A4-origins.pdf.

13. J. Robert Oppenheimer, 'Physics in the contemporary world', *Bulletin of the Atomic Scientists* 4, no. 3 (1 March 1948): 65–86, https://doi.org/10.1080/00963402.1948.11460172.

14. Matteo Valleriani, 'The *Nova scientia*: transcription and translation', 18 April 2013, https://edition-open-sources.org/sources/6/12/index.html.

15. W. J. Hurley and J. S. Finan, 'Military operations research and Digges's Stratioticos', *Military Operations Research* 22, no. 2 (2017): 39–46.

16. Michael Brooks, *The Quantum Astrologer's Handbook* (Scribe, 2017).

17. Calling the side length of the big cube t, Cardano can say that $t^3 = u^3 + (t$

$- u)^3 + 2tu(t - u) + u^2(t - u) + u(t - u)^2$, where u is the side length of one of the small cubes. Rearrange this and you will end up with $(t - u)^3 + 3tu(t - u) = t^3 - u^3$. Then you can simply say that $x = t - u$, and you have exactly the formula you started with: $x^3 + mx = n$, where $m = 3tu$ and $n = t^3 - u^3$. A little more manipulation (start by substituting the fact that $u = m/3t$ into $t^3 - u^3$) and you get you to $(t^3)^2 - n(t^3) - m^3/27 = 0$. You might be thinking this isn't helping — but it is. What you have now is a quadratic equation, with t^3 instead of x, and you already know how to solve those.

18. Phil Patton, 'The shape of Ford's success', *The New York Times*, 24 May 1987, sec. Magazine, https://www.nytimes.com/1987/05/24/magazine/the-shape-of-ford-s-success.html.

19. jdhao, 'The mathematics behind font shapes — Bézier curves and more', 27 November 2018, https://jdhao.github.io/2018/11/27/font_shape_mathematics_bezier_curves/.

20. Tony Rothman, 'Genius and biographers: the fictionalization of Evariste Galois', *American Mathematical Monthly* 89, no. 2 (1982): 84–106, https://doi.org/10.2307/2320923.

21. 'Celebrate the mathematics of Emmy Noether', *Nature* 561, no. 7722 (12 September 2018): 149–50, https://doi.org/10.1038/d41586-018-06658-w.

22. Albert Einstein, 'The late Emmy Noether.; Professor Einstein writes in appreciation of a fellow-mathematician', *The New York Times*, 4 May 1935, https://www.nytimes.com/1935/05/04/archives/the-late-emmy-noether-professor-einstein-writes-in-appreciation-of.html.

23. *The Collected Papers of Albert Einstein, Volume 8: The Berlin Years: Correspondence, 1914-1918* (English Translation Supplement), page 217 (245 of 742)', https://einsteinpapers.press.princeton.edu/vol8-trans/245.

24. F. Hirzebruch, 'Emmy Noether and topology', http://webcache.googleusercontent.com/search?q=cache:iMmQ_GuV370J:www.mathe2.uni-bayreuth.de/axel/papers/hierzebruch:emmy_noether_and_topology.ps.gz+&cd=13&hl=en&ct=clnk&gl=uk.

25. Sergey Brin and Lawrence Page, 'The Anatomy of a Search Engine', http://infolab.stanford.edu/~backrub/google.html.

26. Kurt Bryan and Tanya Leise, 'The $25,000,000,000 eigenvector: the linear algebra behind Google', *SIAM Review* 48, no. 3 (January 2006): 569–81, https://doi.org/10.1137/050623280.

27. P. Wei, L. Chen, and D. Sun, 'Algebraic connectivity maximization of air transportation network: the flight routes' addition/deletion problem', *Transportation Research Part E: Logistics and Transportation Review* 61 (January 2014): 13–27.

28. Harald Hagemann, Vadim Kufenko, and Danila Raskov, 'Game theory modeling for the Cold War on both sides of the Iron Curtain', *History of*

the Human Sciences 29, no. 4–5 (1 October 2016): 99–124, https://doi.org/10.1177/0952695116666012.

29.   'Solving Fermat: Andrew Wiles', https://www.pbs.org/wgbh/nova/proof/wiles.html.

30.   Keith J. Devlin, *The Millennium Problems: The Seven Greatest Unsolved Mathematical Puzzles of Our Time* (New York: Basic Books, 2002).

Chapter 4: Calculus

1.   Gallup Poll, http://ibiblio.org/pha/Gallup/Gallup%201940.htm.

2.   In the July poll, the full question was, 'If the question of the United States going to war against Germany and Italy came up for a national vote within the next two weeks, would you vote to go into the war or to stay out of the war?' In September, the American public was asked 'Which of these two things do you think is the most important for the United States to try to do — to keep out of war ourselves or to help England win, even at the risk of getting into the war?' In December 1940, another poll repeated this question. Sixty per cent of respondents said the United States should help England.

3.   Ralph Ingersoll, *Report on England: November 1940* (New York: Simon and Schuster, 1940), http://archive.org/details/ReportOnEngland.

4.   Peter Reese, 'The showgirl and the Schneider Trophy', The History Press, https://www.thehistorypress.co.uk/articles/the-showgirl-and-the-schneider-trophy/.

5.   Jeffrey Quill, *Spitfire: a test pilot's story* (Manchester: Crécy, 1998).

6.   F.W. Lanchester, *Aerodynamics: constituting the first volume of a complete work on aerial flight* (London: Constable, 1907).

7.   Alfred Price, *Spitfire: a documentary history* (London: Macdonald and Jane's, 1977).

8.   Lance Cole, *Secrets of the Spitfire* (Pen & Sword, 2018).

9.   Stephen T. Ahearn, 'Tolstoy's integration metaphor from *War and Peace*', *American Mathematical Monthly* 112, no. 7 (2005), 631–38.

10.   Roberto Cardil, 'Kepler: The Volume of a Wine Barrel', http://www.matematicasvisuales.com/loci/kepler/doliometry.html.

11.   'A timeline of HIV and AIDS', HIV.gov, 11 May 2016, https://www.hiv.gov/hiv-basics/overview/history/hiv-and-aids-timeline.

12.   Alan S. Perelson, 'Modeling the interaction of the immune system with HIV', in *Mathematical and Statistical Approaches to AIDS Epidemiology*, ed. Carlos Castillo-Chavez, Lecture Notes in Biomathematics (Berlin: Springer, 1989), 350–70, https://doi.org/10.1007/978-3-642-93454-4_17.

13.   David D. Ho et al., 'Rapid turnover of plasma virions and CD4 lymphocytes in HIV-1 infection', *Nature* 373, no. 6510 (January 1995): 123–26, https://doi.org/10.1038/373123a0.

14.   Sarah Loff, 'Katherine Johnson biography', NASA, 22 November 2016,

http://www.nasa.gov/content/katherine-johnson-biography.

15.   'Letter from Newton to John Collins, dated 8 November 1676', The Newton Project, http://www.newtonproject.ox.ac.uk/view/texts/normalized/NATP00272.

16.   Richard S. Westfall, *Never at Rest: a biography of Isaac Newton* (Cambridge: Cambridge University Press, 1980).

17.   William John Greenstreet, *Isaac Newton, 1642–1727: A Memorial Volume Edited for the Mathematical Association* (London: G. Bell, 1927).

18.   Jeanne Peiffer, 'Jacob Bernoulli, teacher and rival of his brother Johann', *Electronic Journal for History of Probability and Statistics* 2/1 (June 2006).

19.   Daniel Bernoulli and Sally Blower, 'An attempt at a new analysis of the mortality caused by smallpox and of the advantages of inoculation to prevent it', *Reviews in Medical Virology* 14, no. 5 (2004): 275–88, https://doi.org/10.1002/rmv.443.

20.   Daniel Bernoulli, 'Exposition of a new theory on the measurement of risk', *Econometrica* 22, no. 1 (1954): 23–36, https://doi.org/10.2307/1909829.

21.   'July 1654: Pascal's letters to Fermat on the "problem of points"', http://www.aps.org/publications/apsnews/200907/physicshistory.cfm.

22.   Erdinç Akyıldırım and Halil Mete Soner, 'A brief history of mathematics in finance', *Borsa Istanbul Review* 14, no. 1 (1 March 2014): 57–63, https://doi.org/10.1016/j.bir.2014.01.002.

23.   Fischer Black and Myron Scholes, 'The pricing of options and corporate liabilities', *Journal of Political Economy* 81, no. 3 (1973): 637–54.

24.   Robert C. Merton, 'On the pricing of corporate debt: the risk structure of interest rates', *Journal of Finance* 29, no. 2 (1974): 449–70, https://doi.org/10.1111/j.1540-6261.1974.tb03058.x.

25.   Jørgen Veisdal, 'The Black-Scholes formula, explained', *Medium*, 4 July 2020, https://medium.com/cantors-paradise/the-black-scholes-formula-explained-9e05b7865d8a.

26.   Richard Stimson, 'Einstein's wing flops', https://wrightstories.com/einsteins-wing-flops/.

27.   *The Collected Papers of Albert Einstein, Volume 6: The Berlin Years: Writings, 1914–1917*, p. 402 (430 of 654)', https://einsteinpapers.press.princeton.edu/vol6-doc/430.

28.   B. S. Shenstone, 'The Lotz method for calculating the aerodynamic characteristics of wings', *Aeronautical Journal* 38, no. 281 (May 1934): 432–44, https://doi.org/10.1017/S036839310010940X.

29.   Price, *Spitfire*.

30.   R.C.J. Howland and B.S. Shenstone, 'I. The inverse method for tapered and twisted wings', *The London, Edinburgh, and Dublin Philosophical Magazine and Journal of Science* 22, no. 145 (1 July 1936): 1–29, https://doi.org/10.1080/14786443608561663.

31.   'Adolf Galland: winged knight of the Luftwaffe', *Warfare History Network* (blog), 12 September 2016, https://warfarehistorynetwork. com/2016/09/12/adolf-galland-winged-knight-of-the-luftwaffe/.

32.   Heinz Knoke and R. J Overy, *I Flew for the Führer: the memoirs of a Luftwaffe fighter pilot* (London: Frontline Books, 2012), http://site.ebrary. com/id/10651960.

Chapter 5: Logarithms

1.   Steinar Thorvaldsen, 'Early numerical analysis in Kepler's new astronomy', *Science in Context* 23, no. 1 (March 2010): 39–63, https://doi.org/10.1017/ S0269889709990238.

2.   Brian Rice, Enrique González-Velasco, and Alexander Corrigan, 'John Napier', in *The Life and Works of John Napier*, ed. Brian Rice, Enrique González-Velasco, and Alexander Corrigan (Cham: Springer, 2017), 1–60, https://doi.org/10.1007/978-3-319-53282-0_1.

3.   kip399, *Arithmetic, Population and Energy — Full Length*, 2012, https:// www.youtube.com/watch?v=sI1C9DyIi_8.

4.   Victor Stango and Jonathan Zinman, 'Exponential growth bias and household finance', *Journal of Finance* 64, no. 6 (2009): 2807–49, https:// doi.org/10.1111/j.1540-6261.2009.01518.x.

5.   Matthew R. Levy and Joshua Tasoff, 'Exponential-growth bias and overconfidence', *Journal of Economic Psychology* 58 (1 February 2017): 1–14, https://doi.org/10.1016/j.joep.2016.11.001.

6.   Alessandro Romano. Chiara Sotis, Goran Dominioni, and Sebastián Guidi, 'The public do not understand logarithmic graphs used to portray COVID-19', *LSE COVID-19* (blog), 19 May 2020, https://blogs.lse.ac.uk/ covid19/2020/05/19/the-public-doesnt-understand-logarithmic-graphs-often-used-to-portray-covid-19/.

7.   Tobias Dantzig and Joseph Mazur, *Number: the language of science* (New York: Plume, 2007).

8.   Kevin Brown, *Reflections on Relativity* (Lulu.com, 2011).

9.   'Henry Briggs — biography', Maths History, https://mathshistory.st-andrews.ac.uk/Biographies/Briggs/.

10.   'Statistical Accounts of Scotland: Killearn, County of Stirling, OSA, Vol. XVI, pp. 108–09, 1795, https://stataccscot.edina.ac.uk/static/statacc/dist/ viewer/osa-vol16-Parish_record_for_Killearn_in_the_county_of_Stirling_ in_volume_16_of_account_1/.

11.   Walter W. Bryant, *A History of Astronomy* (London, Methuen, 1907), http://archive.org/details/ahistoryastrono01bryagoog.

12.   Max Caspar and Clarisse Doris Hellman, *Kepler* (New York: Dover Publications, 1993).

13.   Christopher J. Sangwin, 'Newton's polynomial solver', https://

www.sliderulemuseum.com/REF/NewtonsPolynomialSolver_byChristopherJSangwin2002.pdf.

14. Richard Davis and Ted Hume, *Oughtred Society Slide Rule Reference Manual* (Roseville, CA: The Oughtred Society), http://www.oughtred.org/books/OSSlideRuleReferenceManualrevA.pdf.

15. 'The curve is exponential', https://www.atomicarchive.com/history/first-pile/firstpile_10.html.

16. Claudia Dreifus, 'In the footsteps of his uncle, then his father', *The New York Times*, 14 August 2007, sec. Science, https://www.nytimes.com/2007/08/14/science/14conv.html.

17. U.G. Mitchell and Mary Strain, 'The number e', *Osiris* 1 (1936): 476–96.

18. Académie des inscriptions et belles-lettres (France) Auteur du texte, 'Le Journal Des Sçavans', issue, Gallica (1846): 51, https://gallica.bnf.fr/ark:/12148/bpt6k57253t.

19. Wolfgang Karl Härdle and Annette B. Vogt, 'Ladislaus von Bortkiewicz — statistician, economist and a European intellectual', *International Statistical Review* 83, no. 1 (April 2015): 17–35, https://doi.org/10.1111/insr.12083.

Chapter 6: Imaginary Numbers

1. 'Dudley Craven', http://www.dudleycraven.com/.

2. Paul J. Nahin, *An Imaginary Tale: The Story of $\sqrt{-1}$* (Princeton, NJ: Princeton University Press, 2016).

3. Emelie Kenney, 'Cardano: "arithmetic subtlety" and impossible solutions', *Philosophia Mathematica* s2–4, no. 2 (1 January 1989): 195–216, https://doi.org/10.1093/philmat/s2-4.2.195.

4. Roger Penrose, *The Road to Reality: A Complete Guide to the Laws of the Universe* (London: Random House, 2005).

5. Richard P. Feynman, *The Character of Physical Law* (London: Penguin Books, 1992).

6. Guido Bacciagaluppi and Antony Valentini, *Quantum Theory at the Crossroads: reconsidering the 1927 Solvay Conference* (Cambridge: Cambridge University Press, 2009).

7. Eugene P. Wigner, 'The unreasonable effectiveness of mathematics in the natural sciences. Richard Courant Lecture in Mathematical Sciences delivered at New York University, May 11, 1959', *Communications on Pure and Applied Mathematics* 13, no. 1 (1960): 1–14, https://doi.org/10.1002/cpa.3160130102.

8. John Baez, 'The octonions', *Bulletin of the American Mathematical Society* 39, no. 2 (2002): 145–205, https://doi.org/10.1090/S0273-0979-01-00934-X.

9. Simon L. Altmann, 'Hamilton, Rodrigues, and the quaternion scandal',

Mathematics Magazine 62, no. 5 (1989): 291–308, https://doi.org/10.2307/2689481.

10.   Melanie Bayley, 'Alice's adventures in algebra: Wonderland solved', *New Scientist*, 19 December 2009, https://www.newscientist.com/article/mg20427391-600-alices-adventures-in-algebra-wonderland-solved/.

11.   Melanie Bayley, Email communication with author, 22 April 2020.

12.   Walter Isaacson, *Einstein: his life and universe* (New York: Simon & Schuster, 2007).

13.   Paul Halpern, *Einstein's Dice and Schrödinger's Cat: how two great minds battled quantum randomness to create a unified theory of physics* (New York: Basic Books, 2016).

14.   Graduate Mathematics, *Michael Atiyah, From Quantum Physics to Number Theory [2010]*, 2015, https://www.youtube.com/watch?v=zCCxOE44M_M.

15.   *Proceedings of the International Electrical Congress Held in the City of Chicago, August 21st to 25th, 1893* (New York, American Institute of Electrical Engineers, 1894), http://archive.org/details/proceedingsinte01chicgoog.

16.   'Modern Jove hurls lightning at will; million-horse-power forked tongues crackle and flash in laboratory. To perfect arresters Dr. Steinmetz's artificial bolts shatter wood, and wire vanishes in dust', *The New York Times*, 3 March 1922, https://www.nytimes.com/1922/03/03/archives/modern-jove-hurls-lightning-at-will-millionhorsepower-forked.html.

17.   Letters to the Editor, *LIFE* magazine, May 14, 1965, 27, (Time Inc., 1965).

18.   David Packard, David Kirby, and Karen R. Lewis, *The HP Way: how Bill Hewlett and I built our company* (New York: HarperBusiness, 1995).

Chapter 7: Statistics

1.   Ian Sutherland, 'John Graunt: a tercentenary tribute', *Journal of the Royal Statistical Society, Series A* 126, no. 4 (1963): 537, https://doi.org/10.2307/2982578.

2.   Max Roser, Esteban Ortiz-Ospina, and Hannah Ritchie, 'Life expectancy', Our World in Data, 23 May 2013, https://ourworldindata.org/life-expectancy.

3.   'From the height of this place', Official Google Blog, https://googleblog.blogspot.com/2009/02/from-height-of-this-place.html.

4.   'Timeline of statistics', http://www.statslife.org.uk/images/pdf/timeline-of-statistics.pdf.

5.   Francis Galton, 'Eugenics: its definition, scope and aims', *American Journal of Sociology* 10, no. 1 (July 1904): 45–50, https://galton.org/essays/1900-1911/galton-1904-am-journ-soc-eugenics-scope-aims.htm.

6.   George Bernard Shaw, 'Lecture to the Eugenics Education Society', *Daily Express*, 4 March 1910.

7. Winston Churchill, 'Asquith Papers, MS 12, Folios 224–8', 10 December 1910.

8. Stephen Jay Gould, *The Mismeasure of Man*, revised and expanded (New York: Norton, 1996).

9. Adrian J. Desmond and James R. Moore, *Darwin's Sacred Cause: race, slavery and the quest for human origins* (London: Penguin Books, 2013).

10. Angela Saini, *Superior: the return of race science* (London: 4th Estate, 2020).

11. Francis Galton, 'Vox Populi', *Nature* 75, no. 1949 (7 March 1907): 450–51, https://galton.org/essays/1900-1911/galton-1907-vox-populi.pdf.

12. Francis Galton, 'I. Co-relations and their measurement, chiefly from anthropometric data', *Proceedings of the Royal Society of London* 45, no. 273–279 (1 January 1889): 135–45, https://doi.org/10.1098/rspl.1888.0082.

13. Francis Galton, 'The history of twins' (1875), https://galton.org/essays/1870-1879/galton-1875-history-of-twins.htm.

14. Simon Scarr and Marco Hernandez, 'Drowning in plastic: visualising the world's addiction to plastic bottles', Reuters (4 September 2019), https://graphics.reuters.com/ENVIRONMENT-PLASTIC/0100B275155/index.html.

15. 'The Sick and Wounded Fund', *The Times*, 8 February 1855.

16. Lynn McDonald (ed.) *Florence Nightingale: The Crimean War*, The Collected Works of Florence Nightingale, Vol. 14 (Waterloo, Ontario: Wilfrid Laurier University Press, 2010).

17. Michael D. Maltz, 'From Poisson to the present: applying operations research to problems of crime and justice', *Journal of Quantitative Criminology* 12, no. 1 (1 March 1996): 3–61, https://doi.org/10.1007/BF02354470.

18. World Health Organization, 'Cancer: carcinogenicity of the consumption of red meat and processed meat', accessed 8 January 2021, https://www.who.int/news-room/q-a-detail/cancer-carcinogenicity-of-the-consumption-of-red-meat-and-processed-meat.

19. Ronald Aylmer Fisher et al., *Statistical Methods, Experimental Design, and Scientific Inference* (Oxford [England]; New York: Oxford University Press, 1990).

20. Tommaso Dorigo, 'Demystifying The Five-Sigma Criterion', Science 2.0, 14 August 2014, https://www.science20.com/quantum_diaries_survivor/demystifying_fivesigma_criterion_part_ii-118442.

21. Royal Statistical Society, 'Royal Statistical Society concerned by issues raised in Sally Clark case', news release (23 October 2001), http://www.inference.org.uk/sallyclark/RSS.html.

22. Vincent Scheurer, 'Convicted on Statistics?', Understanding Uncertainty, https://understandinguncertainty.org/node/545.

23. In your mind, the probability of my innocence (H) is 30 per cent, or 0.3.

What we want to work out is the probability of my innocence, given the evidence E. For this we first need to find the quantity $P(E)$, the probability that my blood matches the blood at the crime scene. This is the sum of two factors. The first is the probability it matches if I am innocent *and* that I am innocent:

$$P(E|H) \times P(H)$$

where $P(E|H)$ is the probability of the evidence matching anyone innocent in the population: 35 per cent or 0.35. So this term is 0.35×0.3, which is 0.105.

The second term is the probability the evidence matches if I am not innocent (which is 100 per cent, or 1) *and* I am not innocent, which you have set at 65 per cent, or 0.65:

$$P(E \mid \text{not } H) \times P(\text{not } H)$$

So this is 1×0.65, which is 0.65.

Now we add those two terms together to cover all the possibilities: $0.105 + 0.65 = 0.755$. This is $P(E)$, the probability of my blood matching that at the crime scene. The overall probability that I am innocent, given this evidence, $P(H|E)$, is a combination of your original estimate, the probability of an innocent match with the evidence, and $P(E)$, the probability of my blood matching that found at the scene. It is given by the equation:

$$P(H \mid E) = P(H) \times \frac{P(E \mid H)}{P(E)}$$

$$= 0.3 \times \frac{0.35}{0.755}$$

$$= 0.14$$

That means the evidence suggests you should now put the chances of me being innocent at 14 per cent.

24. 'State *v.* Spann, 617 A.2d 247, 130 N.J. 484', CourtListener, https://www.courtlistener.com/opinion/2389693/state-v-spann/.

25. 'State v. Spann', Casetext, https://casetext.com/case/state-v-spann-17.

26. Thomas Levenson, *Newton and the Counterfeiter: the unknown detective career of the world's greatest scientist* (London: Faber, 2010).

27. E. G. V. Newman, 'The gold metallurgy of Isaac Newton', *Gold Bulletin* 8, no. 3 (1 September 1975): 90–95, https://doi.org/10.1007/BF03215077.

28. Joan Fisher Box, 'Guinness, Gosset, Fisher, and small samples', *Statistical Science* 2, no. 1 (February 1987): 45–52, https://doi.org/10.1214/ss/1177013437.

29. David Brillinger, 'John W. Tukey: his life and professional contributions',

Annals of Statistics 30 (1 December 2002), https://doi.org/10.1214/aos/1043351246.

30. Francis Galton, 'Personal identification and description', *Nature* 38 (21–28 June 1888): 173–77, 201–02, https://galton.org/essays/1880-1889/galton-1888-nature-personal-id.pdf.

31. Simon Newcomb, 'Note on the frequency of use of the different digits in natural numbers', *American Journal of Mathematics* 4, no. 1/4 (1881): 39, https://doi.org/10.2307/2369148.

32. 'From Johnstown flood to research lab — a success story', *The Michigan Alumnus*, 28 October 1939.

33. Frank Benford, 'The law of anomalous numbers', *Proceedings of the American Philosophical Society* 78, no. 4 (1938): 551–72.

Chapter 8: Information Theory

1. Brandon C. Look, 'Gottfried Wilhelm Leibniz', in *The Stanford Encyclopedia of Philosophy*, ed. Edward N. Zalta, Spring 2020 (Metaphysics Research Lab, Stanford University, 2020), https://plato.stanford.edu/archives/spr2020/entries/leibniz/.

2. Jerry M. Lodder, 'Binary arithmetic: from Leibniz to von Neumann', in *Resources for Teaching Discrete Mathematics*, ed. Brian Hopkins (Washington DC: Mathematical Association of America, 2009), 169–78, https://doi.org/10.5948/UPO9780883859742.023.

3. Jan Krikke, *Digital Dragon: the road to Nirvana runs through the Land of Tao* (CreateSpace, 2017).

4. 'Explanation of binary arithmetic (1703)', http://www.leibniz-translations.com/binary.htm.

5. Mary Everest Boole, *Indian Thought and Western Science in the Nineteenth Century* (The Ceylon National Review, 1901), http://archive.org/details/indianthoughtwes00bool.

6. George Boole, *An Investigation of the Laws of Thought on which Are Founded the Mathematical Theories of Logic and Probabilities* (London: Walton and Maberly, 1854).

7. J. Venn, 'I. On the diagrammatic and mechanical representation of propositions and reasonings', *The London, Edinburgh, and Dublin Philosophical Magazine and Journal of Science* 10, no. 59 (1 July 1880): 1–18, https://doi.org/10.1080/14786448008626877.

8. C.E. Shannon, 'A symbolic analysis of relay and switching circuits', *Transactions of the American Institute of Electrical Engineers* 57, no. 12 (December 1938): 713–23, https://doi.org/10.1109/T-AIEE.1938.5057767.

9. Erico Marui Guizzo, 'The essential message: Claude Shannon and the making of information theory' (master's thesis, Massachusetts Institute of Technology, 2003), https://dspace.mit.edu/handle/1721.1/39429.

10. A.M. Turing, 'Intelligent machinery' (National Physics Laboratory, 1948), https://www.npl.co.uk/getattachment/about-us/History/Famous-faces/Alan-Turing/80916595-Intelligent-Machinery.pdf?lang=en-GB.

11. C.E. Shannon, 'A mathematical theory of communication', *Bell System Technical Journal* 27, no. 3 (July 1948): 379–423, https://doi.org/10.1002/j.1538-7305.1948.tb01338.x.

12. M. Mitchell Waldrop, *The Dream Machine: J. C. R. Licklider and the revolution that made computing personal* (New York: Penguin, 2001).

13. R.V.L. Hartley, 'Transmission of information', *Bell System Technical Journal* 7, no. 3 (1928): 535–63, https://doi.org/10.1002/j.1538-7305.1928.tb01236.x.

14. 'Apollo expeditions to the Moon: Chapter 9.6', https://history.nasa.gov/SP-350/ch-9-6.html.

15. Bill Anders, '50 Years after 'Earthrise,' a Christmas Eve message from its photographer', Space.com, https://www.space.com/42848-earthrise-photo-apollo-8-legacy-bill-anders.html.

16. NASA Content Administrator (Brian Dunbar), 'Excerpt from the "Special Message to the Congress on Urgent National Needs"', NASA (7 August 2017), http://www.nasa.gov/vision/space/features/jfk_speech_text.html.

17. L. Baulert, M. Easterling, S.W. Golomb, and A, Vitterbi, 'Coding theory and its applications to communications systems', JPL Technical Report No. 3267 (1961), http://archive.org/details/nasa_techdoc_19630005185.

18. United States Congress House Committee on Science and Astronautics, *1967 NASA Authorization: Hearings, Eighty-Ninth Congress, Second Session, on H. R. 12718 (Superseded by H. R. 14324)* (Washington, DC: US Government Printing Office, 1966).

19. 'Engineering the communications system for Apollo 11 — general dynamics', https://gdmissionsystems.com/space/apollo11.

20. Email to author from NASA STI Information Desk, 'Re: 19770091020 — design philosophy of', 20 August 2020.

21. G.D. Forney, 'Coding and its application in space communications', *IEEE Spectrum* 7, no. 6 (June 1970): 47–58, https://doi.org/10.1109/MSPEC.1970.5213419.

22. Carl Sagan, *Pale Blue Dot: a vision of the human future in space* (New York: Random House, 1994).

23. 'Robert G. Gallager wins the 1999 Harvey Prize', https://wayback.archive-it.org/all/20070417175505/http://www.ee.ucla.edu/~congshen/robert_gallager.pdf.

24. Robert G. Gallager, 'Low-density parity-check codes' (1963), https://web.stanford.edu/class/ee388/papers/ldpc.pdf.

25. Enrico Guizzo, 'Closing in on the perfect code', *IEEE Spectrum*:

Technology, Engineering, and Science News, https://spectrum.ieee.org/computing/software/closing-in-on-the-perfect-code.

26. 'Mars Reconnaissance Orbiter', https://mars.nasa.gov/mars-exploration/missions/mars-reconnaissance-orbiter.

27. Jung Hyun Bae, Ahmed Abotabl, Hsien-Ping Lin, Kee-Bong Song, and Jungwon Lee, 'An overview of channel coding for 5G NR cellular communications', *APSIPA Transactions on Signal and Information Processing* 8 (24 June 2019), https://doi.org/10.1017/ATSIP.2019.10.https://doi.org/10.1017/ATSIP.2019.10.

28. C.E. Shannon, 'Communication theory of secrecy systems', *Bell System Technical Journal* 28, no. 4 (October 1949): 656–715, https://doi.org/10.1002/j.1538-7305.1949.tb00928.x.

29. Albert W. Small, 'The Special Fish Report (1944)', https://www.codesandciphers.org.uk/documents/small/PAGE001.HTM.

30. B. Jack Copeland, *Colossus: The secrets of Bletchley Park's code-breaking computers* (New York: Oxford University Press, 2010).

31. Walter Jr Koenig, 'Final Report on Project C-43' (1944).

32. Tom Espiner, 'GCHQ pioneers on birth of public key crypto', ZDNet, https://www.zdnet.com/article/gchq-pioneers-on-birth-of-public-key-crypto/.

33. The original NSA post is no longer online, but is excerpted and discussed in Neal Koblitz and Alfred J. Menezes, 'A riddle wrapped in an enigma', 2015, https://eprint.iacr.org/2015/1018.

34. John Archibald Wheeler and International Symposium on the Foundations of Quantum Physics, 'Information, Physics, Quantum: The Search for Links' (Tokyo, 1989).

35. Seth Lloyd, *Programming the Universe: a quantum computer scientist takes on the cosmos* (New York: Knopf, 2006).

36. Jimmy Soni and Rob Goodman, *A Mind at Play: how Claude Shannon invented the Information Age* (New York: Simon & Schuster, 2017).

37. Daniel Oberhaus, 'Marvin Minsky on making the "most stupid machine of all"', https://www.vice.com/en/article/vv7enm/marvin-minsky-on-making-the-most-stupid-machine-of-all-artificial-intelligence.

38. E.O. Thorp, 'The invention of the first wearable computer', in *Digest of Papers. Second International Symposium on Wearable Computers (Cat. No.98EX215)*, 1998, 4–8, https://doi.org/10.1109/ISWC.1998.729523.

39. Rogers, 'Claude Shannon's cryptography research during World War II and the mathematical theory of communication', in *1994 Proceedings of IEEE International Carnahan Conference on Security Technology*, 1994, 1–5, https://doi.org/10.1109/CCST.1994.363804.

40. John Horgan, 'Claude Shannon: tinkerer, prankster, and father of information theory', *IEEE Spectrum*: Technology, Engineering, and Science

News (27 April 2016), https://spectrum.ieee.org/tech-history/cyberspace/claude-shannon-tinkerer-prankster-and-father-of-information-theory.

41. C. Shannon, 'The Bandwagon (Edtl.)', *IRE Transactions on Information Theory* 2, no. 1 (March 1956): 3–3, https://doi.org/10.1109/TIT.1956.1056774.

Conclusion

1. Plato, *Timaeus*, https://www.gutenberg.org/files/1572/1572-h/1572-h.htm.

2. George Markowsky, 'Misconceptions about the golden ratio', *College Mathematics Journal* 23, no. 1 (1 January 1992): 2–19, https://doi.org/10.1080/07468342.1992.11973428.

3. Le Corbusier, *Towards a New Architecture* (New York: Dover, 1986).

Index

ABOUT THE AUTHOR

Michael Brooks a science writer with a PhD in quantum physics. He is the author of several books, including *13 Things That Don't Make Sense: The Most Baffling Scientific Mysteries of Our Time* and *The Quantum Astrologer's Handbook,* a 2017 *Daily Telegraph* Book of the Year. He lives in the United Kingdom.